Mad about Physics

Also by Christopher Jargodzki

Christopher P. Jargocki. *Science Braintwisters, Paradoxes, and Fallacies.* New York: Charles Scribner's Sons, 1976.

Christopher P. Jargocki. *More Science Braintwisters and Paradoxes.* New York, Van Nostrand Reinhold Company, 1983.

Also by Franklin Potter

Frank Potter, Charles W. Peck, and David S. Barkley (ed.). *Dynamic Models in Physics: A Workbook of Computer Simulations Using Electronic Spreadsheets: Mechanics.* Simonson & Co., 1989.

Mad about Physics

Braintwisters, Paradoxes, and Curiosities

Christopher Jargodzki
and
Franklin Potter

John Wiley & Sons, Inc.
New York • Chichester • Weinheim • Brisbane • Singapore • Toronto

This book is printed on acid-free paper. ♾

Published by John Wiley & Sons, Inc.
Published simultaneously in Canada

Illustrations on pages 18, 28, 33, 81, 83, 84, 85, 96, and 105 copyright © 2001 by Tina Cash-Walsh

Design and production by Navta Associates, Inc.

Library of Congress Cataloging-in-Publication Data:

Jargodzki, Christopher
Mad about physics: braintwisters, paradoxes, and curiosities / Christopher Jargodzki and Franklin Potter.
p. cm.
ISBN 0-471-56961-5 (pbk. : acid-free paper)
1. Physics--Miscellanea. I. Potter, Frank. II. Title.

QC75 .J37 2000
530—dc21 00-039914

Printed in the United States of America

10 9 8 7 6 5 4 3 2 1

To my late father—Zdzislaw Jargocki

C. J.

To my Mark Keppel High School science teachers
Ms. Hager and Mr. Forrester, who first challenged me
in chemistry and physics to bring out
the joy in doing science

F. P.

Contents

Answers

Preface

This is a book of just under four hundred puzzles about singing snow, hot ice, vanishing elephants, primary colors, Cartesian divers, perpetual motion, aerodynamic lift, smoke rings, foghorns, virtual pitch, singing wineglasses, magnet keepers, levitating mice, superwomen, tightrope walkers, antigravity, strange pendulums, tensegrity structures, jumping fleas, and automobiles—to mention but a few of the many subjects. The questions range over the entire field of naked-eye physics—those phenomena we can observe without specialized research equipment. Three additional chapters delve into the physics of sports, earth sciences, and astronomy. Here the questions might deal with high-jump records, curveballs, golf ball dimples, waves at the beach, lightning and thunder, negative charge of the Earth, meandering rivers, orbit rendezvous, the Moon's trajectory around the Sun, and planetary exploration—again, to mention but a few of the topics.

The puzzles range in difficulty from simple questions (e.g., "Why can you warm your hands by blowing gently, and cool your hands by blowing hard?") to subtle problems requiring more analysis (e.g., "Bricks are stacked so that each brick projects over the brick below without falling. Can the top brick project more than its length beyond the end of the bottom brick?"). Solutions and more than three hundred references are provided and constitute about two-thirds of the book.

As these examples would show, most of the puzzles contain an element of surprise. Indeed, the clash between commonsense conjecture and physical reasoning is the central theme that runs through this volume.

Einstein characterized common sense as the collection of prejudices acquired by age eighteen, and we agree: At least in science, common sense is to be refined and often transcended rather than venerated. The present volume tries to undermine physical preconceptions employing paradoxes (from the Greek *para* and *doxos,* meaning "beyond belief") to create cognitive dissonance. "Though this be madness, yet there is method in't." We believe that far from being merely amusing, paradoxes are uniquely effective in addressing specific deficiencies in understanding (cf. Daniel W. Welch, "Using paradoxes," *American Journal of Physics* 48 [1980]: 629–632). Take, for example, the following question: A man is standing on a bathroom scale. Suddenly he squats with acceleration *a*. Will the scale reading increase or decrease? Many students, guided by common sense, will say the reading will increase because the man is pushing down on the scale while squatting. The correct answer is that the reading will *decrease* because while the man is squatting his center of gravity is being accelerated downward, and hence the normal force exerted by the scale must decrease. In tackling paradoxes like this one the contradiction between gut instinct and physical reasoning will for some students be so painful that they will go to great lengths to escape it even if it means having to learn some physics in the process.

Are these paradoxes real or merely apparent? From the point of view of the standard methods of instruction in physics, the counterintuitive conclusions reached in many puzzles in this book clearly only *seem* paradoxical. The conclusions may be unexpected and sometimes even mind-boggling. Nevertheless, except for a few puzzles involving deliberate fallacies, they follow impeccably from the fundamental laws of physics and can be confirmed experimentally. But perhaps we should honor our sense of unease and dig a little deeper. After all, many concepts in physics are mere useful fictions that aid visualization or simplify calculations. Examples include centrifugal forces, electric and magnetic field lines, magnetic poles, and the conventional sense of the electric current, to name just a few. Useful fictions can be dangerous in that a constant vigilance is required to remember their fictitious character. There is a long-standing debate in physics as to whether certain well-established concepts have outlived their usefulness and should be completely eliminated. Heinrich Hertz, one of the early participants in this debate, for example, suggested that Newtonian mechanics be restated without using "force" as a basic concept. In the introduction to his *Principles of Mechanics,* published in 1899, he wrote, "When these painful contradictions are removed, the question as to the nature of force will not have been answered; but our minds, no longer vexed, will cease to ask illegitimate questions." Ludwig Wittgenstein, who knew this passage virtually word for word, was so impressed that he adopted it as a statement of his aim in philosophy: "In my way of doing philosophy, its whole aim is to give an expression such a form that certain disquietudes disappear."

Paradoxes are an embodiment of such disquietudes and, as a result, have played a dramatic role in the history of physics, often foreshadowing revolutionary developments. The counterintuitive upheavals resulting from relativity theory and quantum mechanics only enhanced the reputation of the paradox as an agent for change. Is physical reality inherently paradoxical (i.e., crazy, to use the vernacular), or do paradoxes arise solely in our description of it and are they a signal to discard the old conceptual framework and adopt a new one? As this is not a book in philosophy, we have the right to be evasive, and instead of answering the question directly, we prefer to end with an anecdote about two great luminaries of twentieth-century physics, Niels Bohr and Wolfgang Pauli. Some decades ago Bohr was in an audience listening to Pauli explain his early attempt at reconciling the theory of relativity and quantum mechanics. Afterward, Bohr stood up and said, "We are all agreed that your theory is absolutely crazy. But what divides us is whether your theory is crazy enough."

Acknowledgments

It is difficult to acknowledge all the individuals who have helped to bring this book into print.

In rough chronological order, Christopher Jargodzki would like to express appreciation to:

Martin Gardner, formerly of *Scientific American,* who started the ball rolling by recommending that Charles Scribner's Sons offer a contract to a fledgling author;

The late Richard Feynman, whose visits to the University of California at Irvine were a constant source of inspiration;

Professors Myron Bander and Meinhard Mayer of UC Irvine; Professors Ronald Aaron, Alan H. Cromer, Stephen Reucroft, and Carl A. Shiffman of Northeastern University in Boston; Professors Dennis Faulk, Michael Foster, John Gieniec, Robert E. Kennedy, Donald D. Miller, Michael H. Powers, James H. Taylor, and Alvin R. Tinsley of Central Missouri State University; Patricia Hubbard and Crystal Stewart of CMSU for expert help with word processing portions of the manuscript; Michael Dornan, who suggested several chapter titles; Cheryl Davis; and Charlotte Cunningham.

Franklin Potter would like to thank physicist Julius Sumner Miller, who always encouraged understanding "the little things that make the world go round," for encouraging me to teach a graduate course at UC Irvine in the 1980s, using physics puzzles of this nature to connect physics to daily life for Ph.D. students. More importantly, my wife, Patricia, and our two sons,

David and Steven, continue to be my inspiration and deserve all the thanks I can give them.

Both authors wish to express appreciation to Kate C. Bradford, editor at John Wiley & Sons, Inc., who kept the faith in this project over the many years it took to complete it.

To the Reader

These puzzles are meant for fun. It is not important how many you can solve. In fact, some of them have challenged physicists for decades and generated a lot of research literature. These include Crooke's radiometer, Feynman's inverse sprinkler, the siphon, and the aerodynamic lift—to mention just a few notorious examples. Such questions are usually placed toward the end of each chapter and are typically distinguished by an asterisk. It would be a rare reader who could come up with detailed solutions to all the puzzles. Indeed, sometimes the reader may need to think a bit even to understand the answer. Including all the steps would easily double the size of the book, and for this we offer no apologies. If a reader finds the puzzles perplexing and intriguing, we have accomplished our objective.

Mad about Physics can be read with profit by anyone who has had some exposure to introductory physics and wants to learn more about its application to real phenomena. Most puzzles are nonmathematical in character and require only a qualitative application of fundamental physical principles. Many physical concepts are defined directly or indirectly in various passages, and the definitions can be found with the aid of the index. But even someone who knows the subject will quickly realize it is by no means easy to apply physics to the real world, and in this sense this is not an elementary book.

More than three hundred follow-up references are provided for interested readers. These are included with only some of the puzzles, typically those that are more controversial and hence have substantial research literature. There was no room to include a more complete list of references. For this we apologize to the authors whose work may have been inadvertently overlooked.

1 Temperature Risin'

THERMAL ENERGY IS AN INTEGRAL COMPONENT of our environment, but we tend to forget its severe limitations on our activities. For example, producing a small temperature change often requires large amounts of energy when compared to changing other energy forms, such as sound energy or translational kinetic energy. The problems in this chapter include demonstrating how ice can exist in boiling water, explaining the drinking bird toy, and determining the best way to cook hamburgers. Remember to consider the ideal situation first and then introduce the necessary complications as you engage yourself in these challenges.

1. Thermos Delight!

You are given three identical thermos bottles, *A*, *B*, and *C*. Thermos *A* contains 1 liter of 80 °C water; thermos *B* contains 1 liter of 20 °C water; and thermos *C* has no water. Empty container *D* fits easily into any thermos and has perfect thermally conducting walls.

You are forbidden to mix the hot water with the cold water. Can you heat the cold water with the aid of container *D* and the hot water so that the final temperature of the cold water will be higher than the final temperature of the hot water?

"You damn sadist," said Mr. Cummings. "You try to make people think."

—Ezra Pound

2. Boiling Water with Boiling Water

Immerse a small container of cool water in a pot of boiling water without letting the waters mix. If one waits long enough, will the cooler water inside the small container come to a boil?

Thermodynamics is easy—I've learned it many times.

—Anonymous

3. Gas and Vapor

Is there a difference between a gas and a vapor?

4. Ice in Boiling Water?

It is possible to show that ice need not melt in boiling water. Nearly fill a test tube with cool water and then

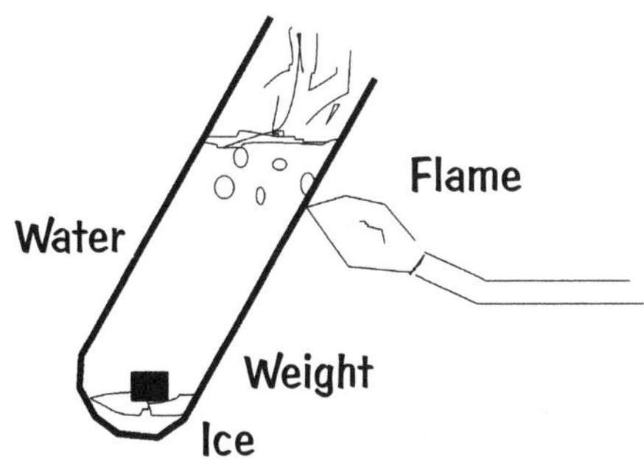

If that this thing we call
  the world
By chance on atoms
  was begot
Which though in cease-
  less motion whirled
Yet weary not
How doth it prove
Thou art so fair and I in
  love?

—John Hall

The energy produced by the atom is a very poor kind of thing. Anyone who expects a source of power from the transformation of these atoms is talking moonshine.

—ERNEST RUTHERFORD (1933)

take a piece of ice and press it down on the bottom of the tube with a small weight. Heat the test tube with a flame that licks only the upper part of the tube. The water will soon be boiling, but the ice at the bottom just won't melt! What is the physics here?

5. Two Mercury Droplets

Two identical mercury droplets at the same temperature combine into one droplet. This final larger droplet is warmer than the original two. Why?

The optimist thinks this is the best of all possible worlds, and the pessimist knows it is.

—J. ROBERT OPPENHEIMER

6. Drinking Bird

The familiar drinking bird dips its beak into the water regularly, then pivots back up to await the next dip. The liquid inside the body and head is methylene chloride, which has a boiling point of 40.1 °C at normal pressure. Unlike the pendulum, the drinking bird does not retain its energy from cycle to cycle but must derive its energy from the surroundings. How does it do so?

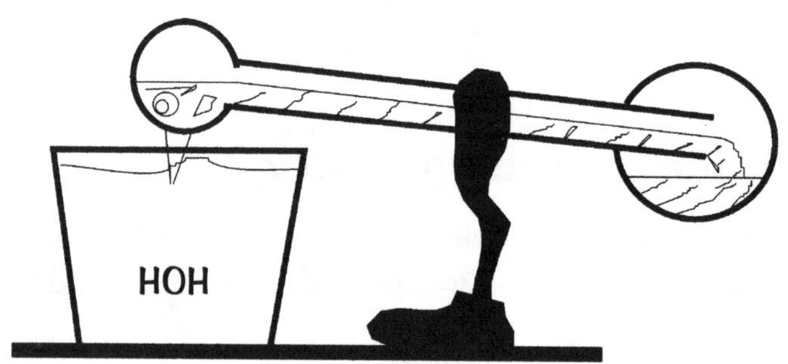

BAKER'S LAW OF INANIMATION

Inanimate objects are scientifically classified into three major categories—those that don't work, those that break down, and those that get lost.

7. Room Heating

If you turn on the heater in your room and after, say, one hour you turn it off, will the total energy of the air in the room be raised by the heating?

8. Shivering at Room Temperature

Room temperature is usually from 18 °C to 22 °C, which is much lower than normal human body temperature of about 37 °C. Shouldn't we be shivering constantly to offset the loss of thermal energy by radiation?

Water freezes at 32 degrees and boils at 212 degrees. There are 180 degrees between freezing and boiling because there are 180 degrees between north and south.

—A STUDENT'S ANSWERS ON A PHYSICS EXAM

9. Identical Spheres Are Heated

Two identical spheres receive identical amounts of thermal energy, the heat transfer occurring so quickly that none is lost to the surroundings. If they begin at the same temperature, but one is on a table and the other is suspended by a string, will the spheres still have the same temperature immediately after the quick addition of thermal energy?

Science answers the question why, and art the question why not.

—SOL LE WITT

10. Cooking Hamburgers

Hamburgers cook faster over a medium flame than over a high flame on a barbeque grill. What do you say?

11. Cooking Hamburgers versus Steak

Why must one cook hamburgers more thoroughly than a slab of steak? After all, both consist of the same meat—beef. And if I like to have my beef cooked rare, what difference can it make whether the meat is a solid slab or ground up?

12. Gasoline Mileage

A gallon of cold gasoline and a gallon of warm gasoline fuel the same car. Which will result in more mileage?

Women often say they feel cold when men feel comfortable because they have more body fat. Body fat accounts for 26 percent of the weight of an average 30-year-old woman compared to just 21 percent in men. Fat is basically inert while muscle fibers are always contracting—even when you seem to be perfectly still—and therefore are generating more heat.

CAYO'S LAW

The only things that start on time are those that you're late for.

—WISDOM FROM THE INTERNET

The first genuine experiment in the freezing of food was conducted by Francis Bacon, who was traveling through snow in 1626 and wondered if it could be used to preserve meat. He stopped his carriage, bought a chicken, had it killed and cleaned. He went outside again, where he stuffed and wrapped the carcass with snow. He wrote that his experiment to preserve flesh "succeeded excellently well," but then, only hours later, he died of pneumonia, brought on by his romp in the snow.

—TOM SHACHTMAN

Great spirits have always encountered violent opposition from mediocre minds.

—ALBERT EINSTEIN

13. Triple Point of Water

What is so special about the triple point of water that the thermodynamic temperature scale is defined by it? That is, the Kelvin is 1/273.16 of the triple point of water. Hint: Consider what happens when a little extra thermal energy enters a closed container that has the ice, water, and water vapor at 0 °C (273.16 K).

14. Cold Salt Mixtures

In 1714, Gabriel Fahrenheit chose as the zero point of his temperature scale the lowest temperature then obtainable in the laboratory: –17.7 °C. This temperature was reached using a mixture of water, granulated ice, and ammonium chloride. Paradoxically, even though the temperature of this mixture drops by almost 18 °C, its energy content remains unchanged if isolated from its surroundings. How is this behavior possible?

15. To Warm, or Not to Warm?

Why can you warm your hands by blowing gently, and cool your hands by blowing hard?

16. Modern Airplane Air Conditioning

Why do newer aircraft on commercial airlines recirculate so much more air than planes did in previous decades?

17. Out! Out! Brief Candle

An inverted glass is placed over a burning candle sitting in a saucer of water. What do you predict? Why?

18. Piston in a Beaker

The illustration shows a beaker of water containing an inverted glass container with a movable piston not touching the water surface. Suppose the water is at room temperature and you raise the piston slowly. What do you predict? Suppose that you begin with the piston initially above boiling water. What do you predict now?

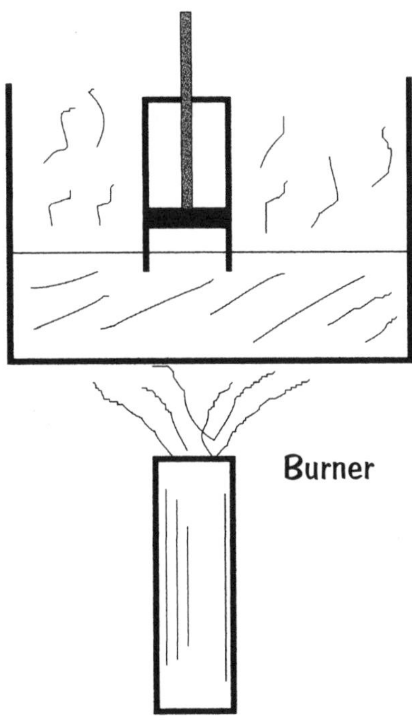

19. Milk in the Coffee

This famous problem is always interesting. Suppose you want to make your morning coffee cool off within five minutes to a more suitable temperature. Do you pour in the cold milk first and then wait five minutes before drinking, or do you wait five minutes before adding the cold milk?

Ludwig Boltzmann, who spent much of his life studying statistical mechanics, died in 1906, by his own hand. Paul Ehrenfest, carrying on the work, died similarly in 1933. [So did another disciple, Percy Bridgman.] Now it is our turn to study statistical mechanics. Perhaps it will be wise to approach the subject cautiously.

—David L. Goodstein

There is not the slightest indication that [nuclear] energy will ever be obtainable. It would mean that the atom would have to be shattered at will.

—Albert Einstein (1932)

In science there is only physics; all the rest is stamp collecting.

—Ernest Rutherford (Ernest Rutherford was awarded a Nobel Prize in Chemistry)

Hell must be isothermal, for otherwise the resident engineers and physical chemists (of which there must be some) could set up a heat engine to run a refrigerator to cool off a portion of their surroundings to any desired temperature.

—HENRY A. BENT

20. Energy Mystery

Two identical lab containers are connected by a narrow tube that has a control valve. Initially, all the liquid is in the left container up to a height h. When the valve is opened, the liquid flows from the left container into the right one, the system finally coming to rest when the levels are $h/2$ in each container.

The initial gravitational potential energy of the liquid is $W\,h/2$, the weight times the height to the center of gravity. In the final state, the total gravitational potential energy is $2\,(W/2)\,(h/4) = W\,h/4$, half of the initial potential energy. Half of the initial gravitational potential energy has disappeared! Why?

A popular cookbook gives directions for determining the power (output) of a microwave oven: Heat a liter of cool water for two minutes. Multiply the temperature change, in Celsius degrees, by 35. The product is the output power in watts.

21. Dehumidifying

Cooling the air in a room by air conditioning should be accompanied by dehumidification. Why?

22. Refrigerator Cooling

Suppose you decide to cool the air in the kitchen by leaving the refrigerator door open. Will this scheme work?

Theoretical physicist: A sadist who thinks up impossible tasks for experimenters.

—IRA M. FREEMAN

23. Air and Water

Air and water at the same temperature, say 25 °C, do not feel the same temperature. One notices this difference immediately upon jumping from 25 °C air into a swimming pool of 25 °C water. Why the difference?

24. Hot and Cold Water Cooling

Two identical wooden pails, without lids, are set out in freezing cold weather. Pail A contains hot water and pail B contains an equal amount of cold water. Which pail will begin to freeze first?

25. Ice-skating on a Very Cold Day

Why is ice-skating harder to do when the temperature of the ice surface is very cold?

26. Singing Snow

Walking on snow on a very cold day can produce a squeak, but there is usually no squeak when the air temperature is barely below freezing. Why not?

27. Contacting All Ice Cubes!

Ice cubes in a bucket often stick together. Why?

28. Hot Ice

Can ice be so hot that your fingers could be burned by contact?

A physicist is an atom's way of knowing about atoms.

—Anonymous

The miracle is that the universe created a part of itself to study the rest of it, and that this part in studying itself finds the rest of the universe in its own natural inner realities.

—John C. Lilly

Every time you drink a glass of water, you are probably imbibing at least one atom that passed through the bladder of Aristotle. A tantalizingly surprising result, but it follows [from the simple] observation that there are many more molecules in a glass of water than there are glasses of water in the sea.

—Richard Dawkins

Unperformed experiments have no results.

—ANONYMOUS

There are a thousand ways to get a problem wrong—not all of them bad—and many ways to get a problem right—not all of them good.

—DAVID GRIFFITHS

Thermodynamics is a funny subject. The first time you go through it, you don't understand it at all. The second time you go through it, you think you understand it, except for one or two points. The third time you go through it, you know you don't understand it, but by that time you are so used to the subject, it doesn't bother you anymore

—ARNOLD SOMMERFELD

29. Walden Pond in Winter

Fish and other organisms appreciate the fact that water expands before freezing. Why would this factoid be important for the fish in Walden Pond in the winter?

30. Lights Off?

Should you diligently turn off incandescent lights at home to "save energy" in winter and in summer? (Quotation marks added to indicate that the phrase actually means less electrical energy is demanded from the energy company.)

31. The Metal Teakettle

Some metal teakettles have metal handles. Isn't this design dangerous?

32. Frozen Laundry

Damp laundry hung out on a line to dry on a cold day will freeze when the temperature gets below freezing. On very cold days, the ice originally on the clothes seems to disappear, but the clothes never get wet. How is this behavior possible?

33. Ice Cream in Milk

Some people like to add milk to a bowl of ice cream. The combination seems much colder to the tongue and the mouth than ice cream alone. What is the physics here?

34. Wearing a Hat in Winter

Why should one wear a hat on a very cold day?

35. Car Parked Outside

A car is parked outside on a clear night and then on a cloudy night. Assuming that the air temperature versus clock reading is the same on both nights, a thick layer of moisture will collect on the car during the clear night but usually not on the cloudy night. What is the reason for this difference?

36. Two Painted Cans of Hot Water

Two cans containing equal amounts of hot water initially at the same temperature are identical except for their color. One can is black on the outside and the other is white on the outside. What do you predict for the water temperatures at later times?

37. Sunshine

Students often wonder how the air can feel cool or even cold in winter even though the sun is shining brightly. Any suggestions?

38. Physicist's Fireplace

Does the amount of thermal energy radiating into a room from a fireplace depend upon how the burning logs are piled?

Science is guided (globally) more by "collective obsessions and controlled schizophrenias" than by disinterested observation.

—Arthur Koestler

George Gamow is reported to have invented a perfect perpetual motion wheel. It used sixes attached to the spokes, which changed into nines as they went over the top to the other side. Thus there was always 50 percent more on one side than on the other. Only the friction in the bearings prevented self-destruction.

Murphy's Law of Thermodynamics
"Things get worse under pressure."

All theory is gray, but the tree of life, my friend, is green.

—JOHANN WOLFGANG VON GOETHE

Many times I have been present at gatherings of people who . . . are thought highly educated and who have with considerable gusto been expressing their incredulity at the illiteracy of scientists. Once or twice I have been provoked and have asked the company how many of them could describe the Second Law of Thermodynamics. The response was cold: it was also negative. Yet I was asking something which is about the scientific equivalent of: "Have you read a work of Shakespeare's?"

—C. P. SNOW

The universe is the externalization of the soul.

—RALPH WALDO EMERSON

39. Blackbody Radiation

The background radiation in the universe is the best blackbody spectrum known. The best approximation to a blackbody radiation source that we can come up with in practice is an oven with a tiny hole. Why does the oven fall short of the ideal?

*40. Uniqueness of Water

Water, silicon, germanium, sterling silver alloys, and lead-tin-antimony alloys have a rare physical property in common: They expand upon freezing. What closely related property is true for water only?

*41. Blowing Hot and Cold

The Ranque-Hilsch vortex tube can separate air into a hot air stream and a cold air stream without any moving parts. Compressed air forced in at room temperature through the side nozzle exits as hot as 200 °C from one arm and as cold as –50 °C from the other arm. There are no heating-cooling devices inside the tube. How does the vortex tube work?

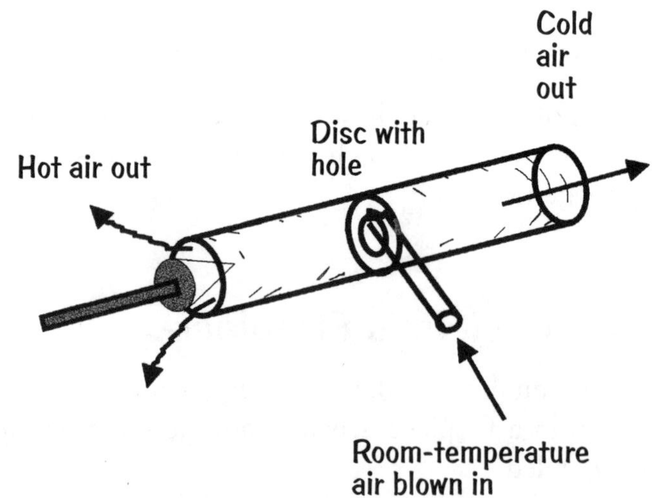

2 Color My World

WE SEE THE WORLD BY THE LIGHT THAT enters our eyes and stimulates nerves, sending messages to our brain. Our tendency is to believe absolutely everything we see. In fact, the old adage "seeing is believing" probably exists in every language and civilization on Earth. Yet, for hundreds of years, experience and optical instruments have been revealing just the opposite—that our eye-brain system is easily fooled. You need only watch an old Western movie and see wagon wheels rotate the wrong way to begin to appreciate our visual gullibility. But there are more subtle effects to contend with, such as floating images and vanishing elephants.

42. Corner Mirrors

Set up two plane mirrors at right angles to each other. Describe your own image as you look into the corner line formed by these two mirrors. Where is your left hand?

But soft! What light through yonder window breaks?

—William Shakespeare

43. The Vanishing Elephant

A magician shows the audience an elephant onstage inside a large cage that has vertical bars and a roof. On cue, the elephant disappears. What is the physics? (Hint: One needs two large plane mirrors.)

We have a paradox in the method of science. The research man may often think and work like an artist, but he has to talk like a book-keeper, in terms of facts, figures, and logical sequences of thought.

—H. D. Smyth

44. Floating Image

Practically everyone has seen the "Mirage Bowl," made from two facing concave mirrors, the top mirror having a hole in its center. The image of a coin or small toy placed inside appears to float in or above the hole. How many reflections are necessary to produce the floating image? Is this a real image or a virtual image?

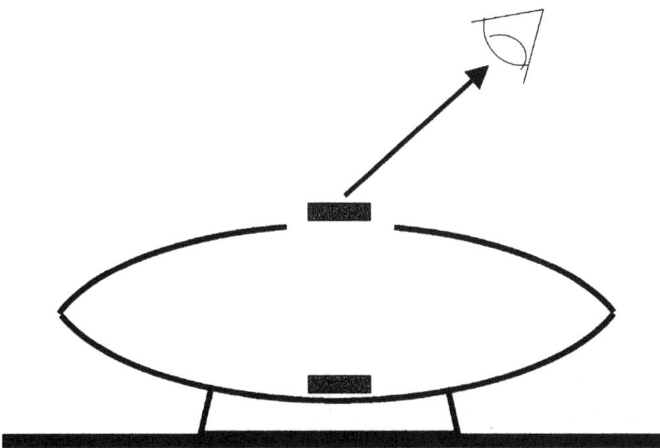

The Sun is 400 times larger across than the Moon and is also roughly 400 times farther away. If it weren't for this remarkable coincidence, there would be no solar eclipses.

The amount of accommodation required when an object is moved from infinity to 1 m is quite small—just enough to shift the image forward 0.06 mm. By contrast, moving the object from 1 m to 1/8 m requires enough accommodation to shift the image an additional 3.51 mm in order to keep it on the retina. Reading this page demands a disproportionate effort. Long study of things too close is one reason for eyestrain; we were apparently designed to look at, and out for, things afar.

—EUGENE HECHT

45. Lighting an Image?

Shine a flashlight beam on the image produced by the "Mirage Bowl" (see the previous question), which has two facing concave mirrors, with the image appearing in the central hole of the top mirror. Will the image be lit by the flashlight?

46. Laser Communicator

You see a space station above the atmosphere and just above the horizon. You desire to communicate with the space station by sending a beam of laser light. You should aim your laser (a) slightly higher than, (b) slightly lower than, (c) directly along the line of sight to the space station.

47. Bent Stick

Turn on a flashlight and shine the beam into a glass tank full of water. The light beam will be seen to sharply change direction *downward* at the point where it enters the water. Now put a straight stick into the water at an angle. The part of the stick in the water will appear to sharply change direction *upward*. Why the contradiction?

Science without religion is lame; religion without science is blind.

—ALBERT EINSTEIN

48. The Pinhole

Can one use a pinhole to measure the diameter of the sun?

49. Window

When you look at an open window from the outside in the daytime, the window appears to be dark. Why?

50. Window Film

Putting a thin metal oxide–coated plastic film on the inside of a glass windowpane to allow less light into a room keeps the room cooler in the summer. Should you remove the film in the winter?

51. Rainbow

In explaining rainbows, one of the first steps is to examine the scattering of the light inside a raindrop. The diagram shows that the light ray enters a spherical raindrop at *A*, undergoes total internal reflection at *B*, and leaves the drop at *C*. At both *A* and *C* there is an air-water interface where refraction occurs to change the ray directions. The light leaving at *C* is split into every color of the visible spectrum, producing the rainbow.

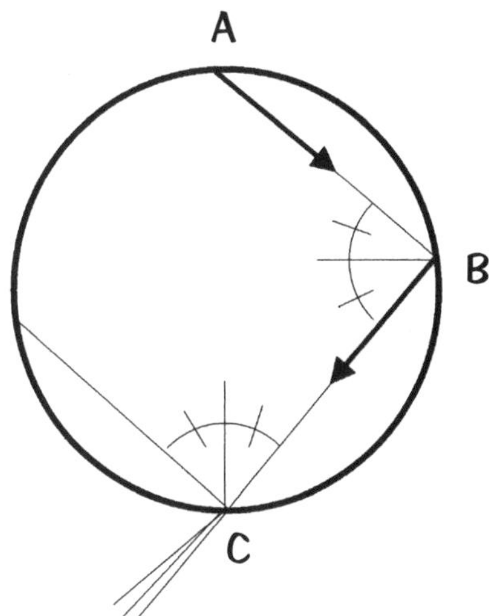

However, one can show that a light ray that has undergone a total internal reflection once inside the drop will never get out of the drop (i.e., that there really would be another total internal reflection at *C*, etc.). How does one resolve this problem to produce a rainbow?

The celestial bodies appear slightly higher above the horizon than they actually are, and this displacement increases the closer they get to the horizon. This accounts for the flattening of the sun on the horizon. At sunset the lower edge of the sun's disc appears, on the average, 35 minutes of arc higher than it actually is, but the top edge only 29. The flattening amounts, therefore, to 6 minutes of arc or about 1/5 of the sun's diameter.

Physics is a form of insight and as such it's a form of art.

—David Bohm

Typically, the air temperature drops by 1 °F per 300 feet of elevation. As a result, distant objects appear a bit lower than they would in the absence of temperature stratification. If instead of dropping, the temperature rises, as is common in the Arctic, then light bends differently and we see a much different world. Distant objects appear raised, bringing normally invisible objects into view.

52. An Optical Puzzle

A rectangular strip of metal foil is bent into the shape of an arc, parallel to the long dimension. If you look at your reflection in the concave surface of this arc, the image will appear inverted. Now rotate the strip slowly through 90 degrees about the line of sight so the long dimension becomes horizontal. What do you see?

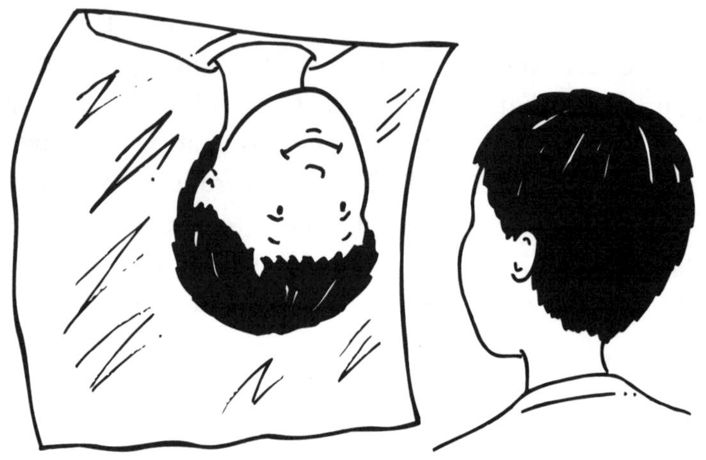

X-rays will prove to be a hoax.

—Lord Kelvin

53. Rearview Mirror

When you flip the lever to tilt the rearview mirror in your automobile to the night position, why doesn't the movement change the direction of view as well as the intensity of the image?

54. Colors

A green blouse looks green because green light is being selectively scattered by the blouse to our eyes. True or false?

55. Primary Colors

The only primary colors of light are red, green, and blue. True or false?

56. Diamond Brilliance

One facet of a brilliant-cut diamond is the entry for a ray of white light having an incident angle several degrees from the normal. After two internal reflections from the back, the ray exits through another, similar facet into your eye. What would you see?

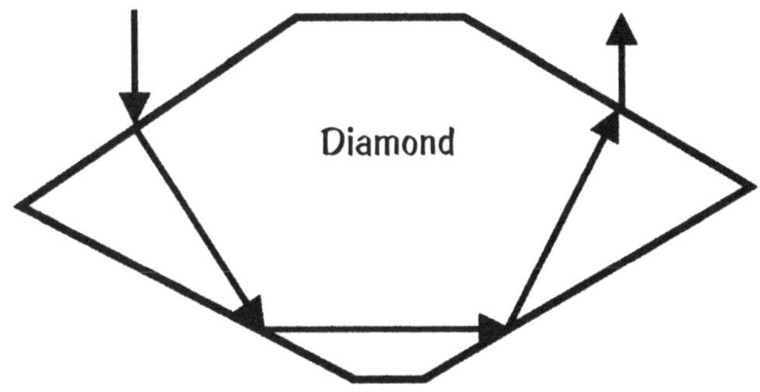

57. White Light Recombined

About a year before Newton studied a prism, in 1665, Francesco Grimaldi was the first to report that a lens could be used to recombine the prism-separated colors of the sun's spectrum to produce white light. What is the geometry needed to accomplish this feat?

58. Prisms

A narrow beam of white light goes through a glass prism that disperses the beam into its constituent colors. Can these rays of color be recombined into white light by transmitting them through an identical, inverted prism?

A TRIPLE PUN

Three brothers bought a cattle ranch and named it "Focus." When their father asked why they chose that name, they replied, "It's the place where the sons raise meat."

—PROFESSOR W. B. PIETENPOL (CONTRIBUTED BY ALBERT A. BARTLETT)

Use the light that is in you to recover your natural clearness of sight.

—LAO TZU

The most incomprehensible thing about the world is that it is at all comprehensible.

—Albert Einstein

59. Squinting

Why does squinting help nearsighted (myopic) people see more clearly? In fact, why not have glasses with pinholes for improving the vision of myopic people?

It is wrong to think that the task of physics is to find out how nature is. Physics concerns what we say about nature.

—Niels Bohr

60. Polarized Sunglasses

Polarized sunglasses are aligned to transmit only vertically polarized light. The reason for this vertical orientation is that most specularly reflected light (the reflected light rays are parallel to each other) from horizontal objects will be horizontally polarized. These sunglasses then reduce the glare of reflections from water, the ground, asphalt, etc. Suppose the right eyeglass is misaligned by about 30 degrees from the vertical. What will you see?

To punish me for my contempt for authority, fate made me an authority myself.

—Albert Einstein

61. Visual Acuity

Diffraction theory tells us that the visual acuity of the human eye should increase inversely with the wavelength—that is, the limiting angle of resolution should be smaller (better) for blues than for reds. The actual human eye has a visual acuity peak in the green at 576 nanometers. Why this discrepancy?

62. Laser Speckle

The reflection of laser light from an object makes it look as if you are seeing the laser spot through stretched-out nylons. How do you explain this observation? When you move your head sideways, the nylon effect moves in the opposite direction!

63. The Red Filter

Write a red *R* and a blue *B* on white paper with crayons. Now look at these letters through a red filter—a material that passes red light only. You do *not* see the *R*! Why not?

Only about 2–5 percent of incandescent lightbulb power is radiated as visible light.

64. Red and Blue Images

Do simultaneous red and blue images on the retina from the same object have the same size?

I hear and I forget
I see and I remember
I do and I understand
—Chinese proverb

65. Colors in Ambient Light

In a room illuminated by the artificial light of lamps and/or fluorescent lights, look at the colors of your clothes. Now walk outside into the sunlight or into the shade and look at your clothes. What will you see? Why?

66. Seeing Around Corners?

Why can we hear but not see around a corner?

67. Stereoscopic Effect

The bright sparkles of light glinting from snow crystals can make them appear to be in the space above the real surface. You can experience this same effect by looking at the surface of a machine-tooled aluminum slab where the marks of the milling bit appear to be almost a foot above the surface! What is the physics here?

Deuterium was discovered in 1932 and its nucleus was at first called a deuton. The story is told, probably apocryphal, that Ernest Rutherford objected to the name deuton—he didn't like its sound. His suggestion was that if his initials, E. R., were inserted it would be all right, and so the deuteron was born.

My suspicion is that the universe is not only queerer than we suppose, but queerer than we can suppose.

—JOHN B. S. HALDANE

68. Eye Color

Why do most newborn human babies have blue eyes? In fact, what physical process gives eyes their color?

69. Metal Clothing

Workers at open-hearth furnaces often wear protective clothing coated on the outside with a thin metal layer. Since metals are excellent conductors of heat, wearing this clothing seems to make no sense (except perhaps in a fashion sense, for the reflections of light present a dazzling display). How does the metal clothing protect the workers from the heat?

It would be possible to describe everything scientifically, but it would make no sense; it would be without meaning, as if you described a Beethoven symphony as a variation of wave pressure.

—ALBERT EINSTEIN

*70. The Sky Should Be Violet

The standard explanation for the reason why the sky is blue invokes Rayleigh scattering, that the blue end of the visible spectrum of sunlight is scattered more effectively than the red end. In fact, this Rayleigh scattering, also known as coherent scattering, is proportional to the fourth power of the light frequency, with the blues being scattered about sixteen times more intensely than the reds. With the blues so effectively scattered into the sky by the air molecules, one sees that the sky is blue and the sun is reddened. But violet light at the very end of the visible spectrum has a greater frequency than the blue light. Why isn't the sky violet instead?

PHYSICS IS LIKE SEX

Sure, it may give some practical results, but that's not why we do it.

—RICHARD FEYNMAN

*71. Crooke's Radiometer I

A radiometer consists of four vanes free to rotate about a pivot inside a glass bulb containing air at very low pressure. One side of each vane is blackened and readily absorbs light; the other side is silvered and reflects most of the incident light. The vanes rotate when they

are illuminated, with the black side moving away from the light source, and the silvered side moving toward the light. The reflective silvered side picks up about twice the momentum per photon reflection as the black side does when absorbing a photon. Why? Because the reflection changes the momentum direction. So why doesn't the radiometer turn the other way?

*72. Crooke's Radiometer II

Light shining on a radiometer makes the vanes rotate in the forward direction—that is, with the black surfaces moving away from the light. Is it possible to make the radiometer rotate in the reverse direction without opening the tube?

*73. Fracto-Emission of Light

Peel adhesive tape from a glass surface in the dark, and if your eyes are dark-adapted, you will see a glow of faint bluish light along the separation line. By the same

There is no such thing as a foolproof device, because fools are so ingenious.

—Anonymous

When a healthy person gazes at a clear blue sky, he will see what looks like little comets darting though the air. Those little stars actually are white blood cells coursing through the eye's retina.

All things are artificial, for nature is the art of God.

—Sir Thomas Browne

physics phenomenon, certain candy wafers when broken in the dark will emit a flash of light. What's the physics here?

*74. Perfect Mirror Reflection

A perfect mirror would reflect incident light of all angles and polarizations, with all the energy going into the reflected rays. Can such a device exist? (Hint: A metallic mirror is omnidirectional but absorbs some of the incident light. And a dielectric mirror made of multiple layers of transparent dielectric materials has an extremely high reflectivity but operates for a limited range of frequencies and within a very narrow range of angles.)

What really interests me is whether God had any choice in the creation of the world.

—Albert Einstein

3 Splish! Splash!

FLUID BEHAVIOR DEPENDS UPON THE PHYSICAL properties of collections of molecules moving together to exert their influence in simple or complex patterns. The challenges in this chapter do not require the consideration of drag forces, but you should not ignore buoyancy, fluid pressure, surface tension, and fluid flow continuity when thinking about the problems of sailing in calm air, drying laundry on a line, and observing double bubbles.

75. Air Has Weight!

A cubic meter of air at sea level weighs how much, approximately? First make an intuitive selection from the list below, then estimate.

(a) Less than 1 ounce	(d) closer to 10 ounces
(b) about 1 ounce	(e) about 1 pound
(c) about 5 ounces	(f) more than 2 pounds

76. Damp Air

A cubic meter of dry air at standard pressure and temperature is weighed on a scale. Then a cubic meter of damp air at the same pressure and temperature is weighed. Which do you predict will weigh more?

77. The Pound of Feathers

Which actually weighs more—a pound of feathers or a pound of iron? "They weigh the same" is not a permitted answer.

78. Sailing in Calm Air!

Suppose you are adrift in a sailboat on a river and the air is absolutely calm everywhere. The river is flowing at 4 knots, but you wish to reach your dock downstream in the least amount of time. Should you raise the sail or leave it down? Will there be any difference?

The normal boiling point of water, 100 °C, has no deep significance for water, since it is defined at normal atmospheric pressure of 1 atm = 1.013×10^5 Pa. In Denver, Colorado, water boils at about 95 °C because of the lower atmospheric pressure. What does have absolute significance is the so-called critical point, which for water is at 647.4 K and 221.2×10^5 Pa (about 218 atm). A gas can be liquefied by compression only if its temperature is below the critical temperature. High-pressure steam boilers in electric-generating plants regularly run at pressures and temperatures well above this critical point.

Some say the world will
end in fire.
Some say in ice
From what I've tasted of
desire
I hold with those who
favor fire.

—Robert Frost

79. The Impossible Dream

Can a sailboat be propelled forward by a fan attached to the deck blowing air into a sail set perpendicular to the centerline of the boat?

80. Lifting Power of a Helium Balloon

The molecular weight of helium gas is 4.0 and the molecular weight of H_2 gas is 2.0. Therefore, a helium balloon of the same volume and pressure has one-half the lifting ability of a hydrogen balloon. Right?

81. Reverse Cartesian Diver

In the traditional Cartesian diver demonstration, you put a medicine dropper or inverted test tube holding trapped air as a diver in a plastic bottle of water and squeeze the opposite sides of the bottle to make the diver go down to the bottom. Take the same bottle of water, and put just a little more water inside the diver so it barely rests on the bottom. The diver can now be made to rise to the top. How is this possible?

82. Cork in a Falling Bucket

A bucket of water containing a cork held down on the bottom by some mechanism is dropped from the top of a building. The cork is released at the moment the bucket is dropped. Where is the cork right before the bucket strikes the ground?

83. Immiscible Liquids

Two immiscible liquids of different densities (such as oil in water) are poured into the bottle shown and shaken vigorously. Initially the resulting mixture is uniformly dispersed. Eventually the liquids separate, with the higher-density liquid settling to the bottom. How does the final pressure on the bottom of the bottle compare with the initial pressure when the liquids were mixed?

84. The Hydrometric Balance

The hydrometric balance consists of a glass tube with a bulb on the immersed end, the whole unit standing vertically in a container of liquid. In the lower bulb end,

Laplace presented Napoleon with a copy of *Celestial Mechanics,* in five volumes. "You have written this huge book," said Napoleon, "without once mentioning the author of the universe." "Sire, I had no need of that *hypothesis,*" replied Laplace.

—Vincent Cronin

The metal gallium will melt in your hand! It is one of the few elements that melt in the vicinity of room temperature. Its normal melting point is 29.8 °C.

liquid or steel balls can be added until the tube is suspended at the desired depth in the surrounding liquid bath.

Suppose this apparatus is on a platform that oscillates up and down, executing simple harmonic motion. What do you predict for the behavior of the hydrometric balance?

85. Child with a Balloon in a Car

Inside a moving automobile, a child holds a helium balloon by a string. All the windows are closed. What will happen to the balloon as the car makes a right turn?

86. The Reservoir behind the Dam

The required strength of a dam is determined by the water behind the dam. Does one need to account for the water in the river that feeds the dam in determining the strength required?

87. Finger in the Water

A bucket of water is placed on one pan of a balance and an equal weight on the other. Will the equilibrium be disturbed if you dip a finger in the water without touching the bucket?

88. The Passenger Rock

A rock is tied to a block of balsa wood, and this system floats on water in a container. When the rock is on the top, exactly half of the wood block is submerged. When the block is turned over so the rock is immersed

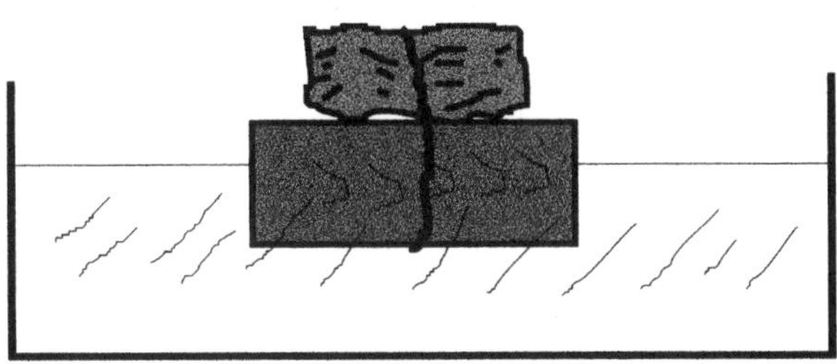

in the water, less than half of the wood block is submerged. Will there be a change in the water level at the side of the container?

89. Archimedes in a Descending Elevator

A block of wood floats in a glass of water. The glass is placed in an elevator. When the elevator starts down with an acceleration $a < g$, will the block stick out higher above the water surface?

90. Three-Hole Can

A water can has three equal-size holes spaced at equal intervals up one side. The water level is maintained at a constant height so that the middle hole is halfway up the water column. The figure shows how the water might flow out of the holes. What do you think about this solution?

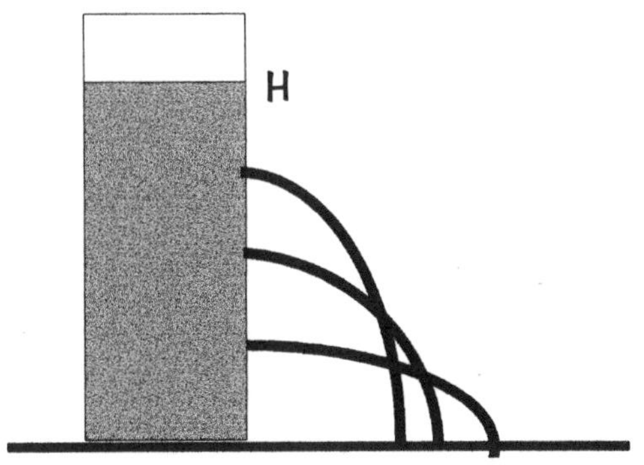

Newton's belief that he was part of the *aurea catena,* the "golden chain" of magi, or unique figures designated by God in each age to receive the ancient Hermetic wisdom, was reinforced by the circumstances of his birth. He was born prematurely, on Christmas day 1642, and was not expected to live. Indeed, that particular parish had a high rate of infant mortality, and Newton later believed that his survival (as well as his escaping the ravages of the plague while still a young man) signified divine intervention. Newton went into extreme rages in his arguments over priority with men such as Hooke and Leibniz, and regarded the system of the world described in the *Principia* as his personal property. He was certain that "God revealed himself to only one prophet in each generation, and this made parallel discoveries improbable." At the bottom of one alchemical notebook Newton inscribed as an anagram of his Latin name, Isaacus Neuutonus, the phrase: *Jeova santus unus*—Jehova the holy one. —Morris Berman

The most beautiful and most profound emotion we can experience is the sensation of the mystical. It is the sower of all true science. He to whom this emotion is a stranger, who can no longer stand rapt in awe, is as good as dead. That deeply emotional conviction of the presence of a superior reasoning power, which is revealed in the incomprehensible universe, forms my idea of God.

—ALBERT EINSTEIN

On being admitted to a hospital, Wolfgang Pauli was told that he was being put into room 137. This number 137, the inverse of the fine-structure constant, particularly puzzled Pauli and continues to challenge physicists today. According to one version of this story, on learning of his room number, Pauli said, "I will never get out of here." The physicist died shortly thereafter.

—F. DAVID PEAT

91. The Laundry Line Revealed!

Why does laundry hung on a clothesline dry from the top down?

92. Pressure Lower than for a Vacuum!

There have been reports that the pressure of sap in trees can get as low as –2MPa—that is, a negative 2 Megapascals! The pressure of a vacuum is defined to be zero; therefore the sap pressure is lower than for a vacuum! Surely the reports must be in error. What do you think?

93. Canoe in a Stream

You are in a canoe without a paddle on a river that is flowing toward a gap between two walls of rock where the stream flows faster. Should you worry that the canoe will enter this constricted region crosswise?

94. Water Flow Dilemma

Two graduated cylinders have a glass tube inside each, the tube inside the cylinder on the left being about

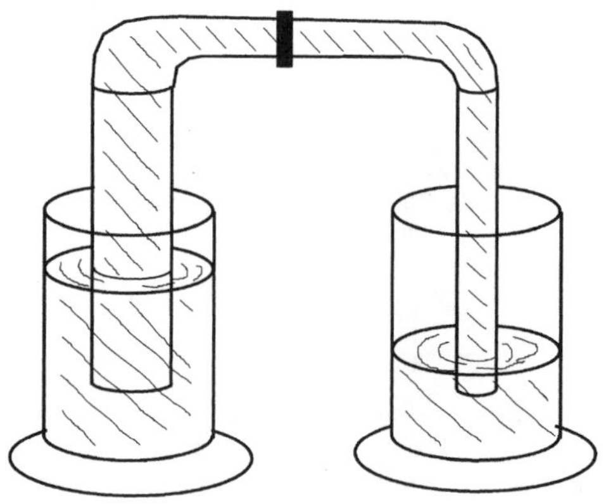

twice the diameter of the other tube. A hose connects the tops of the tubes, and a hose clamp controls the flow between the tubes. The water in the system fills the tubes and the hose as well as most of the left graduated cylinder and a little of the right one. Thus a water height difference exists before the clamp is opened to allow water flow. What do you predict will happen when the clamp is opened?

95. Iron vs. Plastic

A small iron ball and a larger-diameter plastic ball can be arranged to balance each other on a balance. If the balance is inside a bell jar and the air is evacuated quietly (i.e., no convection currents are set up) from the container, what do you predict?

96. Iron in Water

One pan of a balance carries a container with water, and the other a stand with an iron sphere suspended from it (see the diagram). The pans are in balance. Then the stand is turned around so the suspended

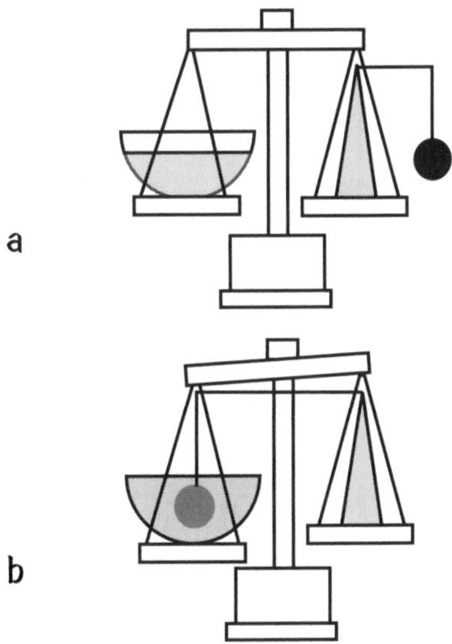

If you're not part of the solution, you're part of the precipitate.

—Steven Wright

The Anthropomorphic Principle

Abstract. We investigate the dynamics of toast tumbling from a table to the floor. Popular opinion is that the final state is usually butter-side down, and constitutes *prima facie* evidence of Murphy's Law ("If it can go wrong, it will"). The orthodox view, in contrast, is that the phenomenon is essentially random, with a 50/50 split of possible outcomes. We show that toast does indeed have an inherent tendency to land butter-side down for a wide range of conditions. Furthermore, we show that this outcome is ultimately ascribable to the values of the fundamental constants. As such, this manifestation of Murphy's Law appears to be an ineluctable feature of our universe.

—The abstract of the paper: Robert A. J. Matthews, "Tumbling toast, Murphy's Law and the fundamental constants," *Eur. J. Phys.* 16 (1995), 172-176.

I am not sure that the mathematician understands this world of ours better than the poet and the mystic. Perhaps it is only that he is better at sums.

—Sir Arthur Eddington

sphere is completely submerged in the water. Obviously the balance is disturbed, since the pan with the stand becomes lighter. What additional weight must be put on this pan to restore equilibrium?

97. Paradox of the Floating Hourglass

An hourglass floats at the top of a closed cylinder that is completely filled with a clear liquid. The cylinder's inside diameter is just large enough to allow the hourglass to move unhindered up and down the tube. When the device is turned over, the hourglass remains at the bottom until about half the sand has fallen into the bottom compartment. The hourglass will then slowly rise to the top. What is the paradox? What is the physics of the operation?

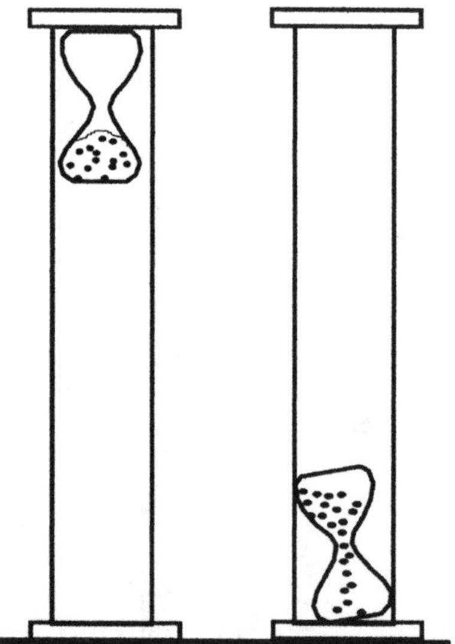

98. Open-Ended Toy Balloon

Inside a large 500-milliliter flask is an inflated balloon. The mouth of the balloon is stretched over the mouth of the flask and its inside is open to the air. Given an identical balloon and flask, could you duplicate this setup?

99. Response of a Cartesian Diver

A minimally buoyant Cartesian diver can be made to behave in interesting ways. A sharp blow with a rubber mallet to the countertop adjacent to the bottle causes the diver to coast to the bottom briefly. What is the physics here?

100. Perpetual Motion

The device shown is a tightly fitted chamber with mercury in the left half and water in the right half. A cylinder is mounted in the center so that it is free to rotate in place along its axis. *One* side of the cylinder—that in the mercury—experiences a greater buoyant force than does the other side, in the water. The difference in torque causes the cylinder to turn clockwise, so one can use the device to drive a generator of electricity. What do you say about this?

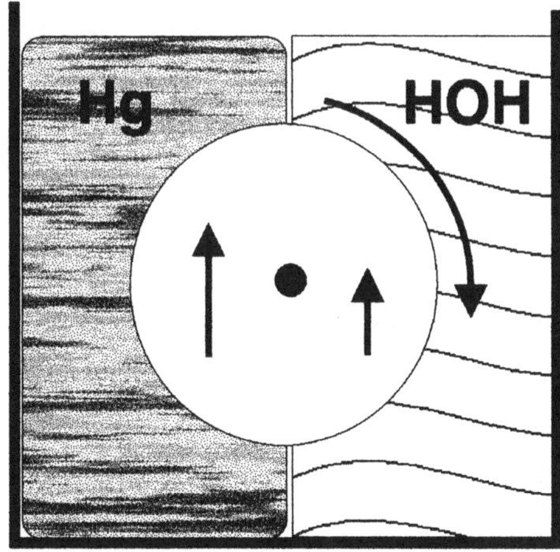

Ninety years ago Duncan McDougall, an American physician, positioned a patient dying from tuberculosis, bed and all, on an enormous beam balance, and waited with scientific curiosity for the end. After several hours, the patient died and "the beam dropped with an audible stroke hitting the lower limiting bar and remaining there with no rebound. The loss was ascertained to be three-fourths of an ounce, as much as a slice of bread."

—Len Fisher

The larger the body of water, the longer its natural wave period will be. The Gulf of Mexico usually has waves spaced about 5 seconds apart, or about 12 waves per minute. The Atlantic Coast has a normal wave period of about 8 waves per minute, and the Pacific has about 7. The height of a wave does not go above 1/7 of its length. If the wave gets too steep, it breaks. As a rule of thumb, the height of a wave will not be greater than half the speed of the wind. A 100 mile-per-hour wind will raise waves with a maximum height of 50 feet.

—GARY LOCKHART

101. Double Bubble

Two soap bubbles (or two identical balloons) are inflated to unequal diameters on opposite ends of a T-shaped tube. The air inlet is then closed, leaving the bubbles connected. What do you predict will happen to the air in the bubbles?

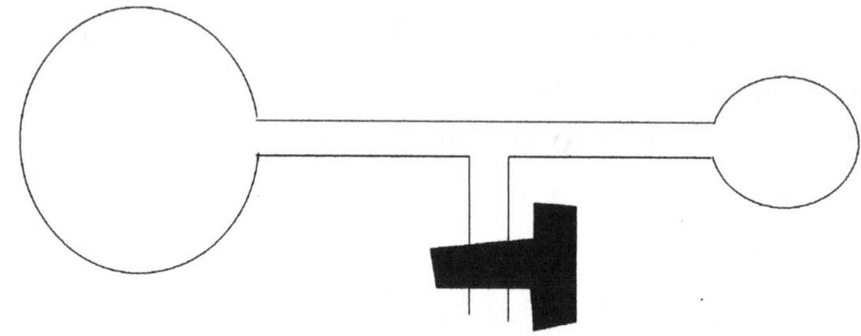

102. The Drinking Straw

When you drink through a straw, the liquid is pushed upward by the ambient air pressure. Suppose now that you simply place the straw upright in the liquid, put your thumb across the top to momentarily seal the top, and lift the straw out from the liquid, still holding it vertically. Lo and behold! The liquid remains in the straw. If you take a scissors or sharp knife and cut a small hole in the section where the liquid is held, what do you predict will happen?

103. Hot-Air Balloon

The explanation most people give for how a hot-air balloon works is that "hot air rises." What do *you* think?

104. Improving the Roman Aqueduct

The Romans built open-topped aqueducts of roughly rectangular cross section to carry water downhill across the countryside from the source to the people. Tall, expensive structures supported them across the valleys. Had the Romans understood fluid flow better, they could have done away with all the tall support structures and built closed tubes on the ground over the hills and down into the valleys and up over the hills again. As long as the water head is above the highest hill elevation along the journey, the water will flow all the way to the destination.

Suppose one of the hills along the way is actually higher than the water source head. Will the closed-tube system still work?

I like relativity and
quantum theories
Because I don't under-
stand them
and they make me feel
as if space shifted
about like a swan
that can't settle,
refusing to sit still and
be measured:
and as if the atom were
an impulsive thing
always changing its
mind.

—D. H. Lawrence

105. Barroom Challenge

Three identical drinking glasses are arranged as shown. Glasses *A* and *B* are filled with water while they are

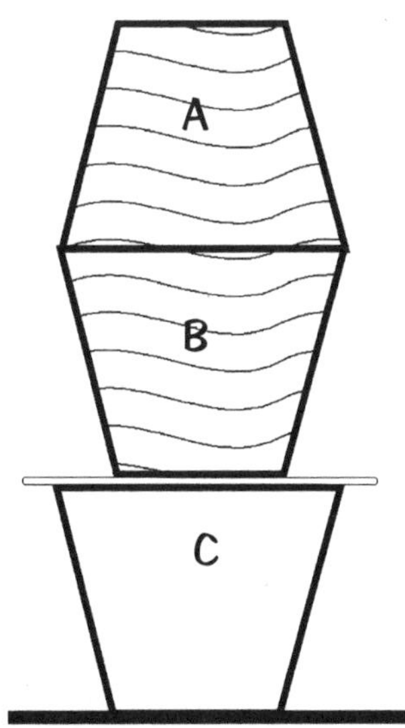

A minus sign is a matter of taste.

—ANONYMOUS

Clocks run "clockwise" because they imitate sundials. In the Northern Hemisphere, the shadows rotate in the direction we now call "clockwise."

submerged and then put together before removing them from the water. Glass *B* is supported on glass *C* by a few hollow stirrers. Additional hollow stirrers are available on the counter. The challenge is to transfer the water (at least most of the water) from glass *A* to glass *C*. *Conditions:* At no time may the participant physically touch or move the glasses or the hollow stirrers supporting glass *B*. The additional stirrers may be moved but cannot touch the glasses or the support stirrers.

106. Tire Pressure

The pressurized air inside an automobile tire supports the weight of the car, right? To check this idea, you would first measure the tire pressure when the tire supports its share of the weight. Then you would jack up the car until the tire no longer touches the road. When you measure its air pressure now, will there be any difference between the two measurements?

*107. The Siphon

Siphons have been used since ancient times to transport liquids up over the edge of a container and into another one at a lower level. Despite its long history, many people consider the operation of a siphon to be a mystery. Some people think that in a siphon the liquid is moved by air pressure. However, a siphon can operate in a vacuum! How do siphons really work? Why isn't a siphon self-starting?

*108. Reverse Sprinkler

When a common rotating lawn sprinkler operates so that water flows out of the nozzles, the rotation direction is opposite to the motion of the exiting water. In the inverse mode, the sprinkler is submerged in a bath of water, and water is pushed into the nozzles. What do you predict for the rotation direction now?

I don't pretend to undersand the Universe—it's a great deal bigger than I am.

—Thomas Carlyle

*109. Spouting Water Droplets

If you slide a styrofoam cup filled with water across a finished wooden surface at about 10 cm/s, you can see droplets of water shoot upward to about 20 cm. Why does this spouting occur?

4 Fly like an Eagle

THE DRAG FORCES CAN ACT IN DIVERSE WAYS TO complicate the motion when a solid object moves in a fluid or when a fluid flows in its container. In some cases turbulence results, and a simple first approximation to explain the behavior often progresses into a supercomputer model that is almost beyond comprehension. So begin your adventure into these challenges with a simple approach, lest your investigation rapidly escalate in difficulty.

110. Vertical Round Trip

An object is thrown straight up. When the effects of the air are ignored, the total time for the round trip up and down can be calculated if the initial velocity upward is known. But suppose that the atmospheric air effects are taken into account. The round-trip time will be longer, right, because the object is moving slower? Does the elapsed time for the downward journey match the elapsed time for the upward journey?

Subtle is the Lord, but malicious He is not.

—Albert Einstein

111. It's a Long Way to the Ground!

A sphere falls through the air. Does the terminal velocity value depend upon the altitude of the drop? Can a sphere dropped later from the same altitude pass the original sphere?

When asked by a colleague what he meant by that, he replied: "Nature hides her secret because of her essential loftiness, but not by means of ruse."

—Abraham Pais

112. Galileo's Challenge Revised

A person simultaneously drops a bowling ball and a much lighter plastic ball of the same diameter from the same height in the air. What do you predict?

113. Falling Objects Paradox

Consider two identical spherical objects. One is dropped from rest from height H above the ground. The other is fired off horizontally from the same height H at the same instant. Do they both hit the ground at the same time? Suppose the earth's curvature is considered. What happens now?

"There will be no heavy duties," Wolfgang Pauli told the physicist who was to serve as his assistant. "Your job is, every time I say something, contradict me with the strongest arguments."

—Barbara Lovett Cline

114. Iceboat

Can an iceboat (which has runners) travel faster on the ice than the speed of the wind that propels it?

The Wright brothers built and flew gliders. By the time they were in their early thirties they had made over one thousand glider flights. The particular challenge that intrigued them was how to power a glider with the recently-invented internal combustion engine. All available motor-car engines were too heavy, so in 1903 the Wrights built a lighter engine. It had four horizontal cylinders of 4-inch bore and 4-inch stroke, and developed some 12 horse-power. They fitted it on a linen-covered wooden glider, so that it would turn two airscrews in opposite directions.

—VINCENT CRONIN

115. The Flettner Rotor Ship

The Flettner rotor ship has a large vertical cylinder along the midline that can be rotated in either direction by a small engine. Suppose the ship is headed west and the wind comes directly from the south. Which direction should the cylinder rotate for the ship to move forward, or does the rotation direction matter at all?

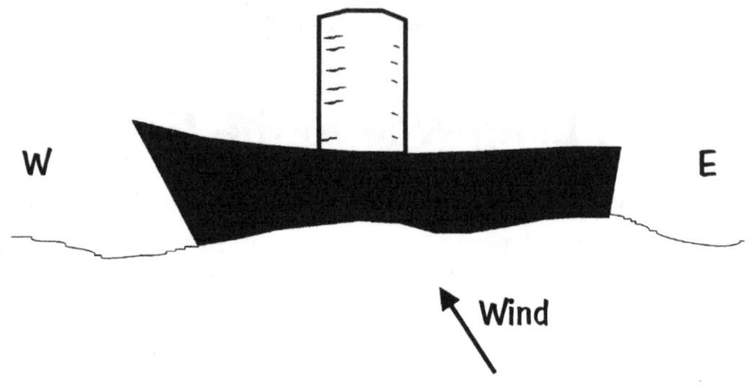

116. The Lift Force Is Greater, Isn't It?

When an airplane is not accelerating in any direction but is flying in a straight line at a constant altitude at a constant speed, the lift force balances the weight and the thrust balances the drag. When the plane enters a stable, constant rate of climb, the lift force is greater than the weight of the plane. Or is it?

117. Floating Rafts

Watching rafts floating down a river, we see those close to the center floating faster than those near the banks. Also, heavily loaded rafts float faster than lightly loaded ones. Why?

118. Dubuat's Paradox

Suppose that you hold a stick in a stream flowing with speed *V*. Then you tow the same stick in the same orientation with the speed *V* through still water. As long as the relative velocity is the same, one might conclude that the water resistance would be the same in both cases. Is it?

I have not the smallest molecule of faith in aerial navigation other than ballooning or of the expectation of good results from any of the trials we hear of.

—Lord Kelvin (1896)

119. Airfoil Shapes in the Airstream

Compare an airfoil facing the airstream with its rounded edge (a) with the same airfoil turned around so its knifelike edge is facing the airstream (b). In which position will the airfoil offer less resistance?

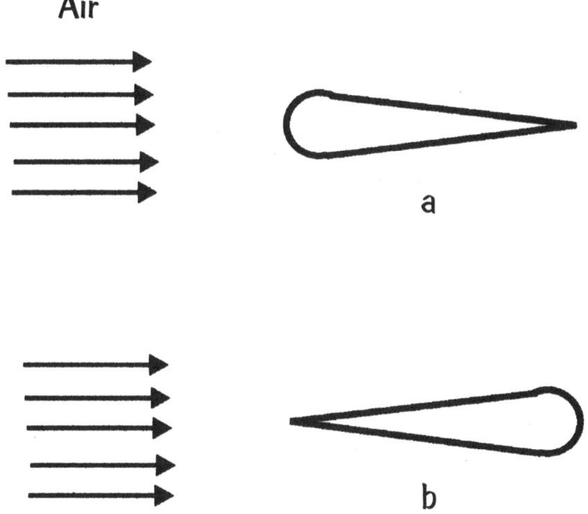

In science, the credit goes to the man who convinces the world, not to the man to whom the idea occurs.

—Sir Francis Darwin

120. Airfoil Shapes in the Waterstream

Compare an airfoil with its rounded edge facing the waterstream (a) with the same airfoil turned around so

The air drag on a 1.0 ft^2 perpendicular area is about 25 lb at 100 mi/h, and it increases to 100 lb at 200 mi/h.

We're number (1/2) $\log_{10}100$

—Bumper Sticker

that its knifelike edge is facing the waterstream (b). In which position will the airfoil offer less resistance?

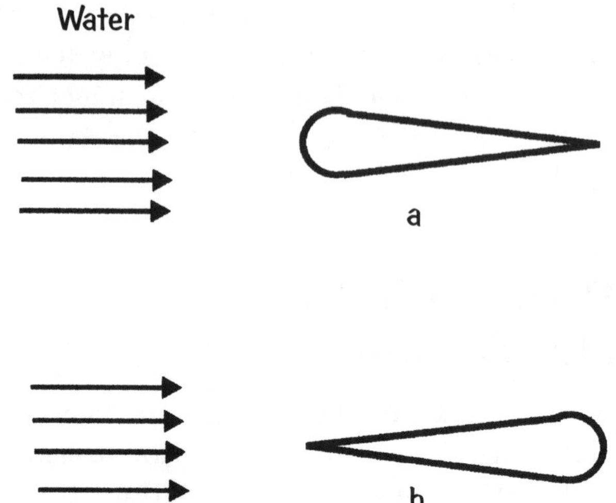

121. Wire vs. Airfoil

The drawing shows an airfoil 10 inches thick at its thickest part and a round wire 1 inch in diameter. Which shape will produce less drag in the same airflow?

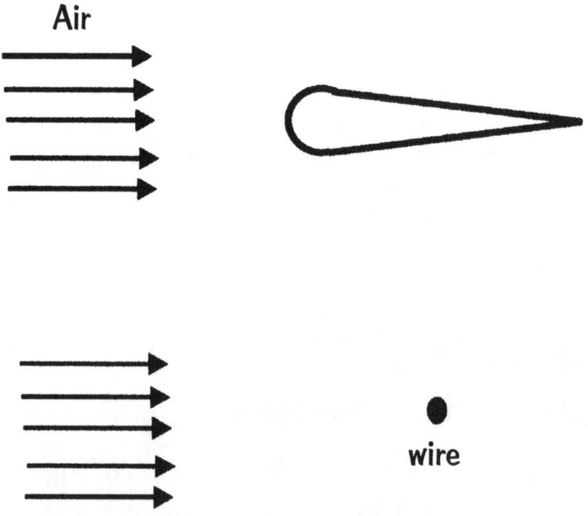

122. Hole-y Wings!

In the past decade or so, some airplane wings have been perforated with millions of regularly spaced tiny holes by the airplane manufacturer. These holes are about one-thousandth of an inch in diameter and are laser-made. Why would such a costly process be undertaken?

Most people gain weight in winter. The effect is so regular that airlines take it into account when estimating the fuel they need for flights.

123. Frisbee Frolics

The thrown spinning Frisbee has a front edge that provides lift, but analysis reveals that the trailing part experiences significant downward airflow, so its lift is reduced. Consequently the center of lift is ahead of the center of gravity for the total flight. Is there any problem with this condition?

Research is to see what everyone else has seen but to think what no one else has thought.

—Albert Szent-Gyorgyi

124. Aerobie Frolics

The Aerobie is a thin ring that can be thrown like a Frisbee for distances exceeding 300 meters. Why can the Aerobie be thrown about twice as far as a Frisbee?

125. Kites I

The bridle on some kites has a light spring or a rubber band on the lower line. Why?

So young and so already unknown.

—Wolfgang Pauli about another physicist

126. Kites II

The tail of a kite helps provide lateral stability by pulling opposite to any momentary lateral motion. The tail can be a strip of cloth, a length of paper or plastic, or a cup or several cups. How can these cups, called drogues, maintain their orientation with the "cylindrical" axis parallel to the wind?

An inscription on the wall of a rest room at Princeton University: 2 + 2 = 5 for large values of 2

127. Parachutes

Why do parachutes have at least one vent hole?

128. Strange Behavior of a Mixture

A very interesting liquid can be made by adding cornstarch to vegetable oil at about a one to three or a one to two ratio by volume and mixing well to a light gravy consistency. You can pour the mixture from a glass in a steady stream. Now electrically charge a piece of styrofoam and bring it near to the pouring stream to stop the flow. What is the physics here?

I hesitate to use the word "God," but in my studies of the universe I have come to the conclusion that there is some purpose to it. The universe has organized itself in such a way, as to become aware of itself. As conscious beings, we are part of that purpose.

—Paul Davies

129. Catsup

Catsup from a bottle initially flows slowly but then speeds up, sometimes quite dramatically. Any thoughts about the physics?

130. Coiled Garden Hose

A coiled garden hose can behave in a very strange way: If you pour water through a funnel into the upper end of the coiled hose, none will come out the other end. Even more surprising, very little water will enter the hose. Why?

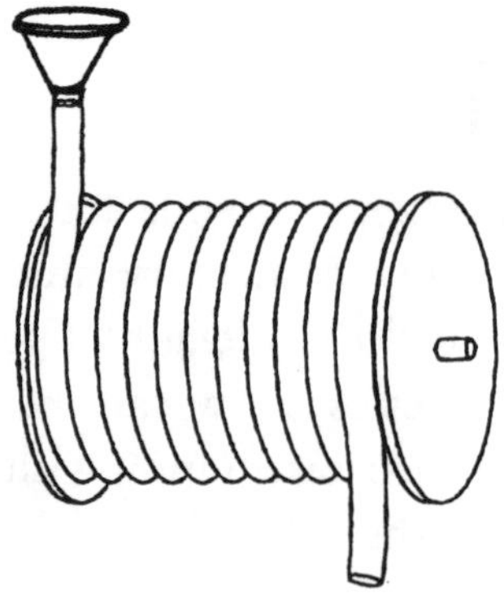

131. Flow from a Tube

Water effluxing from a tube or a hose that is facing downward decreases in diameter into a tapering stream. But a non-Newtonian fluid with its long-chain molecules may behave differently. What do you predict would happen to a non-Newtonian fluid exiting the same orifice?

132. Spheres in a Viscous Newtonian Liquid

Imagine dropping two identical spheres, one after the other, from the same height and at the same spot just above the surface, into a viscous Newtonian liquid. What do you predict about their movement through the liquid?

133. Spheres in a Viscous Non-Newtonian Liquid

Imagine dropping two identical spheres, one after the other, from the same height and at the same spot just above the surface into a viscous non-Newtonian fluid. What do you predict about their movement through the liquid?

134. Animalcules in R < 10-4

Many animalcules (tiny animals with length < 10 microns) live in an environment with a Reynolds number $R < 10^{-4}$! (The Reynolds number is the ratio of inertial force to viscous force.) They move about quite readily. One example is the famous bacterium *E. coli*, which is a cylindrical beastie with a tail flagellum for

I believe that ideas such as absolute certitude, absolute exactness, final truth, etc., are figments of the imagination which should not be admissible in any field of science. . . . This loosening of thinking seems to me to be the greatest blessing which modern science has given us. For the belief in a single truth and in being the possessor thereof is the root cause of all evil in the world.

—MAX BOEN

The North Celestial Pole is less than 1 degree of arc from Polaris. Because of the precession of the earth's axis the NCP will move to within 0.5 degree of arc of Polaris about the year 2012 and then move away again. Around 500 B.C. the NCP was 12 degrees away from Polaris, and it has been close to it only in the last 500 years or so.

All things by immortal
power,
Near or far,
Hiddenly
To each other linked are,
That thou canst not stir
a flower
Without troubling a star.

—FRANCIS THOMPSON

propulsion. None of these animalcules practices reciprocal motion—that is, changing the body into a certain shape and then slowly reversing the sequence to return to the initial shape. Why not?

*135. Lift without Bernoulli

Most explanations for the lift created by airflow past the wings of an airplane invoke the Bernoulli effect. Yet one can explain the origin of lift without using the Bernoulli effect at all. Indeed, the fact that some planes can fly upside down seems to contradict the Bernoulli explanation. How can one explain flight without Bernoulli?

An expert is a man who has made all the mistakes which can be made in a very narrow field.

—NIELS BOHR

*136. Storm in a Teacup

Stir a cup of tea with loose tea leaves in the liquid and most of the tea leaves end up on the bottom in the center. Why?

*137. Smoke Rings I

A smoke ring in still air travels slowly in the direction perpendicular to the plane of the ring (see page 51). In such a ring the smoke particles rotate around the hollow toroidal axis of the doughnut (toroid) in the directions indicated with the arrows. What makes smoke rings travel through the air? Which way will the smoke ring in the diagram travel?

For in much wisdom is
much grief,
And he who increases
knowledge,
increases sorrow.

—Ecclesiastes 1:18

If we knew what it was we were doing, it would not be called research, would it?

—Albert Einstein

*138. Smoke Rings II

Two smoke rings can chase one another, the trailing ring accelerating and shrinking while the leading ring slows down and expands. The smaller ring catches up with the larger one and passes through. Then the roles are reversed and the process is repeated! A fascinating show, but how do we explain it?

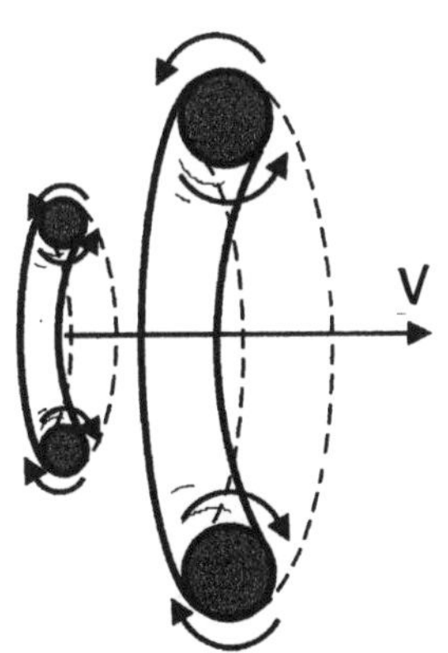

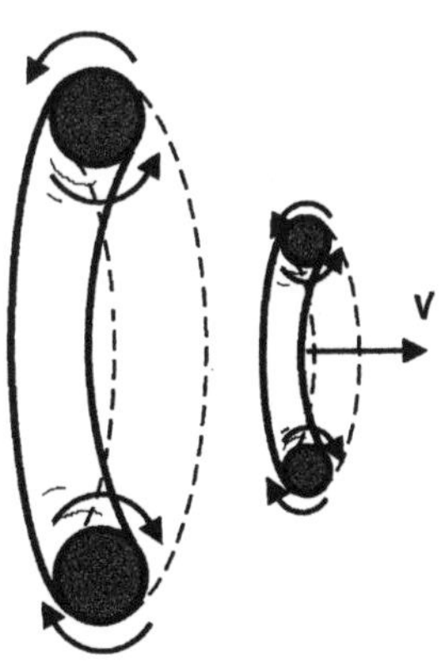

5 Good Vibrations

WE USUALLY REFER TO SOUND IN TERMS OF how it is perceived by human ears. The source of a sound is a vibrating object, and the air transmits the sound waves to our ears, which act with the brain to deliver the information. But sound is important in other aspects of life. Moving a meter stick, for example, requires molecular interactions—a sound wave—to let the other end know that it should follow along. The following challenges barely scratch the surface in the vast realm of acoustics.

139. Conch Shell

Put a conch shell up to your ear and you can hear a marvelous symphony of sounds. Why are these sounds in the conch shell?

140. Hearing Oneself

Most of us will swear upon hearing our own recorded voice that this recorded voice sounds different from the voice we know. Are we victims of an illusion, or is the difference real?

141. A Rumble in the Ears

In a quiet room, put both thumbs in your ears and listen carefully to the low rumbling sound at about 25 hertz or slightly lower in frequency. What is producing this sound?

142. Sound in a Tube

How does a sound wave traveling down a tube get reflected from its open end, from nothing?

143. Those Summer Nights

Why does sound carry well over water, especially in summer?

Kepler's chief source of inspiration was the Pythagorean doctrine of celestial harmony, which he had encountered in Plato. "As our eyes are framed for astronomy, so our ears are framed for the movements of harmony," Plato wrote, "and these two sciences are sisters, as the Pythagoreans say and we agree." In the final book of the *Republic,* Plato portrays with great beauty a voyage into space, where the motion of each planet is attended to by a Siren singing "one sound, one note, and from all the eight there was a concord of a single harmony."

—TIMOTHY FERRIS

The tuning of a six-foot grand piano rises 0.3 percent in frequency (0.05 semitone) on the average when the relative humidity increases by 10 percent.

In her reminiscences Marcia Davenport tells of musical parties given by her mother, the great soprano, Alma Gluck: Sometimes Efrem Zimbalist sat in, or Jascha Heifetz if he was there; and sometimes on the other hand, the Quartet was confronted by Professor Einstein, timidly pleading to be allowed to play second fiddle. He was a ghastly violinist but they stood in awe of his eminence and were honored to be in his company. But their sidelong glances of agony as he scratched and squealed along with them, beaming under his halo of fuzzy white hair! He seldom counted correctly, which confounded them in the world's greatest mathematical physicist.

—FROM *TOO STRONG FOR FANTASY* BY MARCIA DAVENPORT

Music is a secret arithmetical exercise and the person who indulges in it does not realize that he is manipulating numbers.

—G. W. VON LEIBNIZ

144. Cannon Fire

On February 2, 1901, cannons were fired in London to mourn the death of Queen Victoria. The sounds were heard throughout the city but not in the surrounding countryside. Strangely enough, the cannon fire was clearly heard by astonished villagers ninety miles away. How could the sound hop over the outskirts of London and come down ninety miles away?

145. Speaking into the Wind

Why is it difficult to hear upwind from a source of sound, apart from the masking effect of the noise produced by the wind? Is it because the wind "blows" the sound back?

146. Foghorns

Why are foghorns designed to emit only low-pitched sounds?

147. Yodelers' Delight!

How is it that mountain climbers and balloonists often hear and understand persons on the ground, even from a half mile up, while the latter cannot hear and understand them at all? For example, a yodeler on the ground can be well appreciated by those above, but quite often the reverse is not true!

148. Tuning Fork Crescendo

A tuning fork vibrates oppositely at the two prongs and up and down along its handle. If you hold a tuning fork vertically close to one ear and slowly rotate it about the vertical axis passing through the handle, you will hear the sound grow louder and softer.

If interference of the sound waves produced by the two prongs is *not* the cause, what is? (Note: The prongs are about 2 to 3 centimeters apart compared to the sound wavelength of nearly a meter.)

149. Hark!

Can a female speaker usually fill a room with her voice more easily than a male speaker?

150. Rubber and Lead

Sound generally travels much faster in liquids and solids than in gases. Examples include a sound velocity in steel of about 5,000 meters per second; about 1,500 meters per second in seawater; around 340 meters per second in air. Then why is the sound velocity only 1,200 miles per second in lead and, even more surprisingly, only 62 meters per second in rubber?

151. Helium Speech

Why do people's voices sound higher-pitched when they inhale helium?

152. Maestro, Music Please!

The diagrams on page 58 show two concert hall designs differing only in the shape of the ceiling above the orchestra. The numbers show the time differences in milliseconds between the times of arrival of the direct and the reflected sound. Which concert hall will have better acoustics, all other factors being the same?

The great tragedy of science—the slaying of a beautiful hypothesis by an ugly fact.

—Thomas Henry Huxley

The upper range of human air conduction hearing is believed to be no higher than about 24,000 hertz. However, humans can hear well into the ultrasonic range when the ultrasonic stimuli are delivered by bone conduction.

The acoustic energy emitted by 50,000 football fans roaring at the top of their lungs in the course of the game will be barely enough to warm a cup of coffee.

On this account Pythagoras kept a lyre with him to make music before going to sleep and upon waking, in order always to imbue his soul with its divine quality.

—Censorius

Sound travels great distances in arctic regions. Animals in these areas have small ears for they can hear a barking dog up to 15 miles away. In hot desert areas where sound travels poorly, animals have large sensitive ears to maximize sound reception.

—GARY LOCKHART

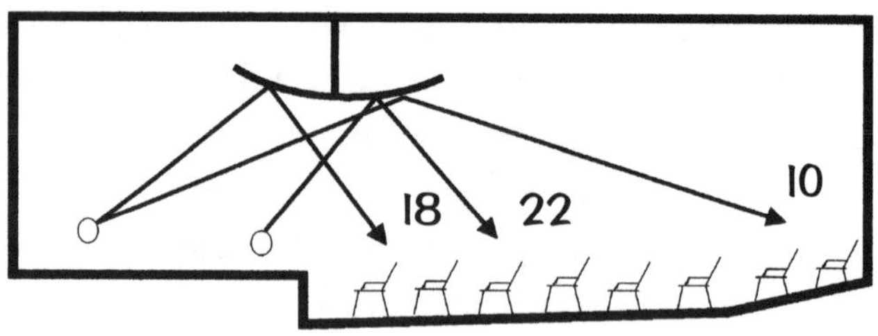

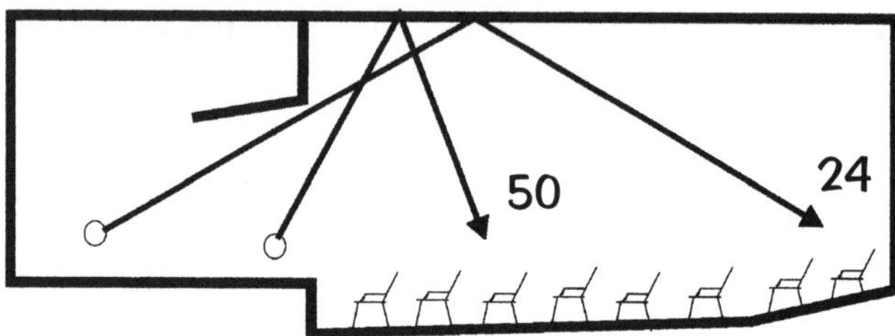

M. E. Bryan and I. Colger (*Acustica* 29 [1973]: 228–233) conclude that as far as intellectual tasks are concerned, noise is a great leveller of performance. In their experiments, the more intelligent subjects suffered a deterioration of performance in a noisy environment, while the less intelligent actually improved.

—THOMAS D. ROSSING

153. The Mouse That Roared

We have all heard the phrase "the mouse that roared," but is there any scientific basis for this feat? On the other hand, could an elephant emit a high-frequency squeak?

154. Bass Notes Galore

How is it possible for telephones and other devices with small speakers to reproduce bass tones that have wavelengths more than ten times their size?

155. Virtual Pitch

Certain types of choral music seem to contain tones that do not exist in the voices of the singers, yet one certainly hears those additional tones. In the music performed by Tibetan monks one often hears the extra tone, sometimes called virtual pitch. What is happening here?

156. Singing in the Shower

In the shower, even a bad singer's voice can occasionally sound beautiful. Any thoughts about this transformation?

157. Scratching Wood

Take a long piece of wood and put your ear to one end. Stretch out your arm and scratch the most distant place on the wood that you can reach. Your scratching will sound quite loud, yet if you take your ear away and go on scratching as before, there is hardly a sound to hear. Why?

158. Simple String Telephone Line

Which orientation of the plastic cup receiver in this string telephone line produces the loudest sounds to the ear?

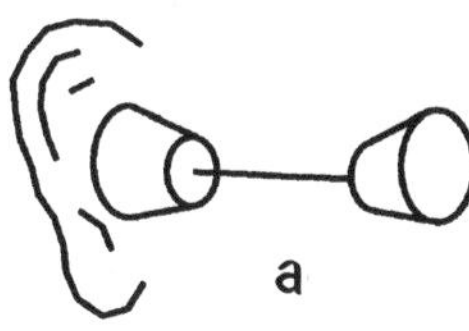

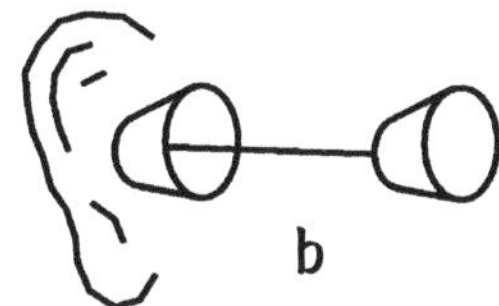

The nasal echo location system of a dolphin is of such a wavelength that it can see through the bodies of other animals and people. Skin, muscle, and fat are almost transparent to dolphins, so they "see" a thin outline of the body, but the bones, teeth, and gas-filled cavities are clearly apparent. The latter include the sinuses, lungs, mouth cavity, bronchi, intestines, etc. Although their echos carry information about the physical condition of another dolphin, a person, or other animal at which they are looking, we do not know, at present, whether dolphins know the significance of what they "see" inside their bodies, or ours. However they do receive the physical evidence of cancers, tumors, strokes, heart attacks, and even emotional states.

—George Barnes

The scientist does not study nature because it is useful; he studies it because he delights in it, and he delights in it because it is beautiful. If nature were not beautiful, it would not be worth knowing, and if nature were not worth knowing, life would not be worth living.

—Henri Poincaré

One might characterize the physical sciences as those for which explicit connection to actual experience constitutes an annoying constraint from which they are trying to liberate themselves more and more.

—Max Delbrück

159. Supersonic Aircraft

Why does an airplane traveling faster than the speed of sound produce two sonic booms?

160. Slinky Sound Off!

Tape one end of a Slinky to a wall (or to a window) to serve as a sounding board and hold the other end in your hand about ten feet from the wall. The Slinky should be suspended, with the neighboring turns not quite in contact with one another. The Slinky is then under a lower than normal operating tension. Now tap the end near your hand with a pencil and listen. What will you hear? (Hint: The velocity at which the sound energy travels is proportional to the square root of the frequency.)

161. Wineglass Singing I

By rubbing the edge of a wineglass with a moist finger, you can make the wineglass sing. If instead you tap that edge with a spoon, what do you predict? Are the two sounds related?

162. Wineglass Singing II

By rubbing the edge of a wineglass with a moist finger, you can make the wineglass sing. What happens to the frequency of the sound when you add water to the singing wineglass?

163. Bell-Ringing Basics

Why are bells of different notes almost always struck in succession and not struck simultaneously?

164. Forest Echoes

Certain locations produce echoes dramatically different in frequencies from the original sounds. For example, if you speak into a large stand of fir trees from a distance, the echo of your voice might come back *raised by an octave*. What is the physics here?

165. Bass Boost

When recorded sound is played back softly with a stereo sound system, why must the bass be given a considerable boost to maintain tonal balance? Why is the bass boost not necessary for the same music played loudly?

166. Personal Attention-Getter

Suppose you are in a crowded store or shopping mall and you would like to be able to send a private sound message to one person standing in the middle of the crowded area. Can this be done? Remember, voice sound waves have wavelengths greater than one meter, so that a normal speaker source will emit sound waves that spread out angularly into the entire room.

167. Musical Staircase

Suppose a musician repeatedly plays the same sequence of notes that move up in an octave. What will many listeners hear? Instead of hearing the pattern stop and then start again, many listeners will hear the pattern ascend endlessly in pitch! Playing the reverse sequence of notes repeatedly, the same listeners hear the pattern descend endlessly. What could be the explanation?

Researchers at the Center for the Neurobiology of Learning and Memory at the University of California at Irvine determined that 10 minutes of listening to Mozart's Sonata for Two Pianos in D Major raised the scores on the standard I.Q. spatial reasoning tasks. Unfortunately, the I.Q. boost dissipated within 10–15 minutes.

—F. H. Rauscher, G. L. Shaw, and K. N. Ky

When Krakatoa, a volcano on a small island in Indonesia, blew up in 1883, its eruptions were heard from 3,000 miles away.

Perfection of means and confusion of ends seems to characterize our age.

—Albert Einstein

168. Where Does the Energy Go?

When two sound waves meet in a region and are canceled by destructive interference, where does the energy go?

*169. A Bell Ringing in a Bell Jar

In a well-known demonstration, an electric bell is hung inside a bell jar. When the air is being pumped out, the bell sounds become less audible until at a pressure of about 1,000 N/m^2 the bell is completely inaudible. Why is this demonstration spurious?

*170. A Well-Tuned Piano

On a well-tuned piano, despite its pleasing sound, each note is actually slightly out of tune with the others. In fact, if the piano were tuned perfectly, its sounds would grate on your ear and in places seem drastically out of tune. What is the explanation?

*171. Driving Tent Stakes into the Ground

A steel tent stake can be driven into hard ground easily and will fit snugly, while an identically shaped wooden stake is hard to drive in and will end up fitting loosely. Why such a dramatic difference?

Certain Tibetan monks can produce the illusion of singing two notes simultaneously. Singing the note whose frequency is about 63 Hz, the first formant frequency is adjusted to about 315 Hz and the second to about 630 Hz. These correspond to the fifth and tenth harmonics of 63 Hz. Because every harmonic in the overtone series of the frequency of 63 Hz is present, that note is heard. The emphasis of 315 and 630 Hz creates the illusion that a note of 315-Hz frequency is also being sung simultaneously. The fifth harmonic of any note is up a musical interval of two octaves and one major third, so the combination of the two notes makes a pleasant-sounding, though unusual-sounding, chord.

—Richard E. Berg and David G. Stork

*172. Loudness

The human ear responds to sound intensity in a non-linear way. Sound intensity is measured with a logarithmic scale called dB. The reference sound intensity level $I_0 = 10^{-12}$ watt/m^2, so that the logarithmic sound intensity level $L_I = 10 \log (I/I_0)$. Therefore, to double the logarithmic sound intensity level, one makes $\log (I_2/I_1) = 2$. Yet, doubling this logarithmic sound intensity level does not usually double the subjective sensation of loudness. Why not?

Eighty percent of problems are trivial, 19.5 percent unsolvable, and genius is required to find and solve the remaining 0.5 percent.

—Hermann Bondi

6 Opposites Attract

WE LIVE IN AN ELECTRICAL WORLD, AND electric and magnetic phenomena permeate all aspects of our daily lives, from neural transmission of signals to and from the brain to computer operation. The challenges in this chapter, such as making a potato battery, analyzing a magnetic sphere, and considering a levitating mouse, have been selected to sharpen your appreciation of one of nature's fundamental interactions. In most cases you should first consider the electrical and magnetic devices to behave in the ideal manner.

173. Three-Bulb Circuit

Three identical lamps are connected in series as shown. When a wire is connected between points A and B, what happens to the brightness of lamps 1, 2, and 3?

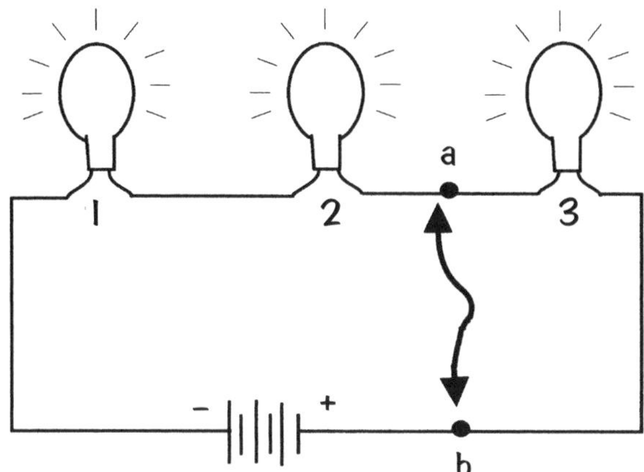

What led me more or less directly to the special theory of relativity was the conviction that the electromotive force acting on a body in motion in a magnetic field was nothing else but an electric field.

—Albert Einstein

174. Potato Battery

You can make a potato battery with copper and zinc (or copper and magnesium) electrodes inserted into the potato. The terminal voltage will be more than 1 volt. What do you predict will happen when you connect a small flashlight bulb across the terminals?

Rubbing will usually build up charge on an ordinary-sized object with a density of up to 10 nanocoulombs per square centimeter, which is a practical limit beyond which there will tend to be discharging into the air.

—Eugene Hecht

175. Resistor Networks

Which of the circuits shown demands more current from the battery?

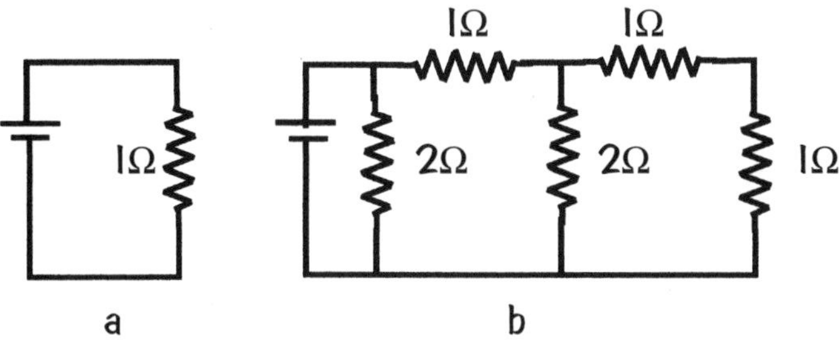

We start off confused and end up confused on a higher level.

—Alan Chalmers ("What is this thing called Science?")

I've never seen electricity, so I don't pay for it. I write right on the bill, "I'm sorry, I haven't seen it all month."

—Steven Wright

176. A Real Capacitor

An isolated, charged capacitor does not keep its charge forever. Why not?

Lectures which *really* teach will never be popular; lectures which are popular will never really teach.

—Michael Faraday

177. Capacitor Paradox

Two identical ideal capacitors each have infinite internal resistance. Charge capacitor *A* to value *Q* and leave capacitor *B* uncharged. Connect the two capacitors by an ideal conductor of zero resistance, and the charges will oscillate between the two capacitors. Suppose that the connecting wire resistance is *R*, with perfect insulation. One observes that the wire becomes hot. What is the source of the energy for the heating? If the wire resistance *R* is actually zero, will the oscillations go on forever?

If all of mathematics disappeared, physics would be set back by exactly one week.

—Richard Feynman

178. Charge Shielding

If you are inside a hollow conductor, you are completely shielded electrically from any outside charges. But what if you reverse this situation by putting a metal shield around a charge? By Gauss's law, you will detect an electric field from the charge inside the metal shield. Is there any way to prevent the electric field from this electric charge from reaching you?

179. Three Spheres

Three identical metal spheres are arranged as shown. Each sphere has a 10-centimeter radius, with sphere centers located at 0 centimeters, 50 centimeters, and 100 centimeters. Sphere *A* is connected to sphere *B*, and sphere *B* is connected to sphere *C* by extremely small-diameter wires (neglect any electrical charge on

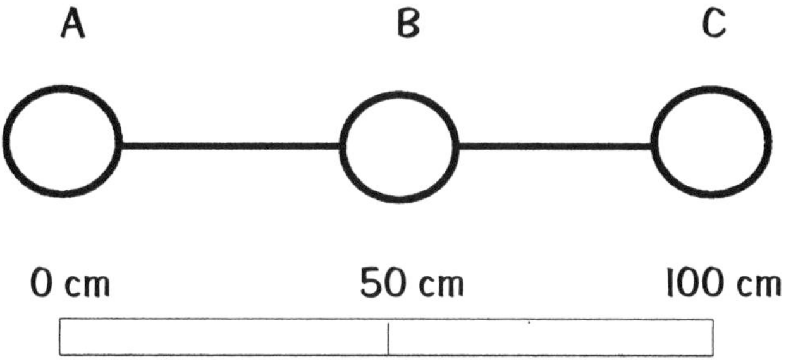

the connecting wires). The total charge on all three spheres is Q. What is the amount of charge on the central sphere?

180. Inductive Charges?

Begin with a neutral electroscope. Is it possible to leave the electroscope with a net positive charge if the only nonneutral object at hand is a negatively charged rod?

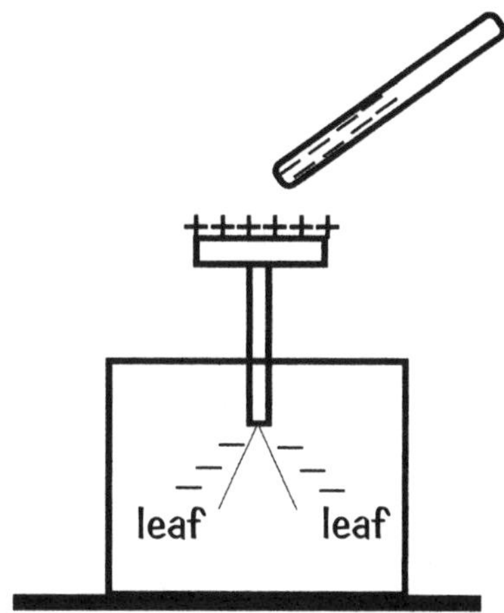

181. Parallel Currents I

Two parallel wires with electrical currents in the same direction experience an attractive force. But an observer moving with a speed of only a few millimeters per second can match the average drift velocity of the

Q: What do you do when your resistors get too hot?
A: Open the switch and coulomb off.

—ANONYMOUS

The obscure we see eventually. The completely obvious, it seems, takes longer.

—EDWARD R. MURROW

Dry skin is a good insulator, with a resistance of 10^6 ohms or more, but wet skin is a good conductor. The resistance of the whole body is about 1,000 ohms.

The capacitance of a human body (measured while standing on 5 cm of insulation) is roughly from 100 pF to 110 pF (for people 68 kg to 105 kg, respectively).

—EUGENE HECHT

electrons in the wires and see the magnetic field of both electron streams completely disappear. For this moving observer, do the wires attract one another?

182. Parallel Currents II

An observer in coordinate frame *S* sees two identical charges at rest. These two charges repel each other via Coulomb's law. These same two charges appear to be parallel currents attracting each other when observed in a coordinate frame S' moving perpendicular to the line joining the charges. Can this paradox be resolved?

183. Rotating Wheel

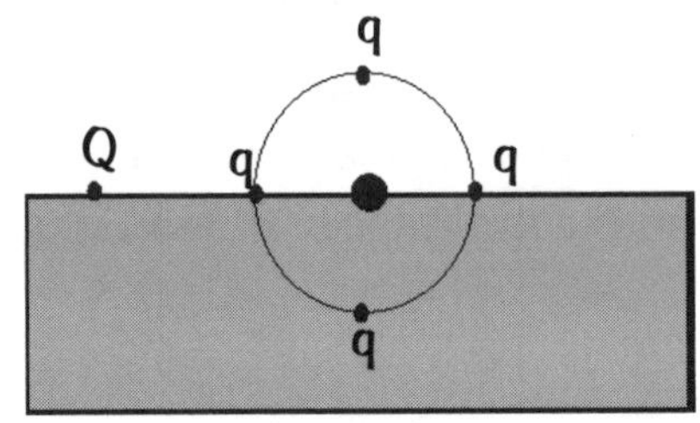

The axle of a wheel lies on the surface of an oil bath, and four equal charges *q* are spaced equally around the perimeter. The fixed charge *Q* repels all four charges *q*. What do you predict will happen to the wheel?

184. Charge Trajectory

A small ideal positive test charge *q* is released from rest at the location shown between two charges +*Q* and –*Q* fixed in position. Will the test charge follow the curve of the electric field line to the –*Q* charge?

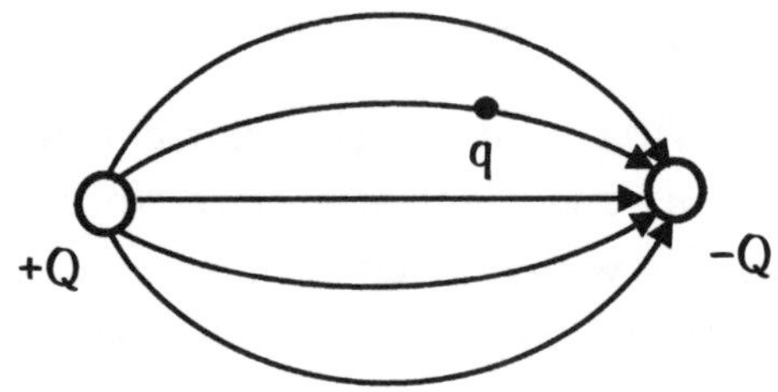

185. Voltmeter Reading

In the circuit shown, a 1-amp current flows through each of the 2-ohm resistors, and the current through the 4-ohm resistor is zero. What will a voltmeter read when connected between points *A* and *B*?

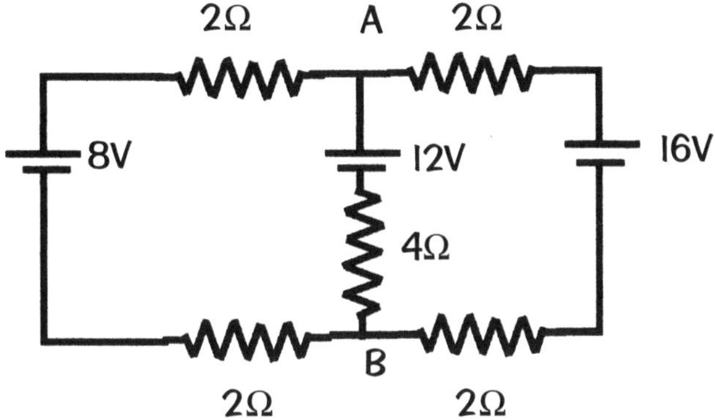

Scientists have odious manners, except when you prop up their theory; then you can borrow money of them.

—Mark Twain

Eminent British politician: "What good is electricity?"
Michael Faraday: "I don't know yet, but someday you'll tax it!"

186. Power Transfer Enigma

Consider the simple circuit with a load resistor *R* connected to a battery with internal resistance *r*. At one extreme, when $R<<r$, the battery acts as an ideal current source. The ideal voltage source occurs when $R>>r$. Under what condition will there be a maximum power transfer? If the efficiency of power transfer approaches 100 percent for the ideal voltage source, why doesn't the ideal voltage source transfer maximum power?

Deep in the human unconscious is a pervasive need for a logical universe that makes sense. But the real universe is always one step beyond logic.

—Frank Herbert

187. Linear Resistance

Doubling the voltage across a standard resistor doubles the current. Right?

The words *anode* and *cathode* were suggested by William Whewell, then Master of Trinity College in Cambridge, in an exchange of letters with Michael Faraday. Faraday hypothesized a giant electrolysis cell accounting for the earth's magnetic field in which the positive electrode would be at the eastern side and the negative electrode at the western side. Whewell recommended to name the electrodes *anode* and *cathode* meaning "the upward way" or "the way from which the sun rises"—that is, east—and "the downward way" or "the way toward which the sun sets"—the west.

—H. GEORGE HAMMON III

Physics is becoming too difficult for the physicists.

—DAVID HILBERT

188. Radioactive Currents

An isolated radioactive source emits alpha particles in all directions. This flow of charge from the source is an electrical current. What magnetic field is associated with this current?

189. Which Is the Magnet?

The only difference between two steel bars is that one is a permanent magnet and the other is unmagnetized. Without using any equipment, how can you tell which is which?

190. Why the Keeper?

Some permanent magnets have a bar called a "keeper" connecting their poles. Why is the keeper important?

191. The Magnet

Close a strong horseshoe magnet with an iron bar *A* as shown. The magnet is strong enough to keep this bar in place. Then take a bar *B* made of soft iron and place it on the magnet as shown. Bar *A* immediately drops off. When bar *B* is removed, the magnet easily holds bar *A* up again. What is the physics here?

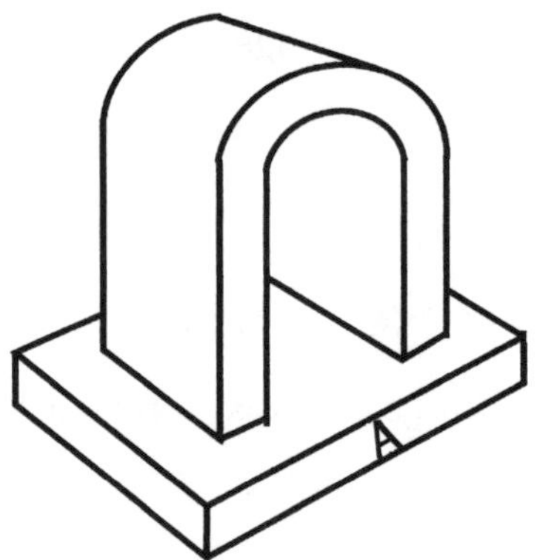

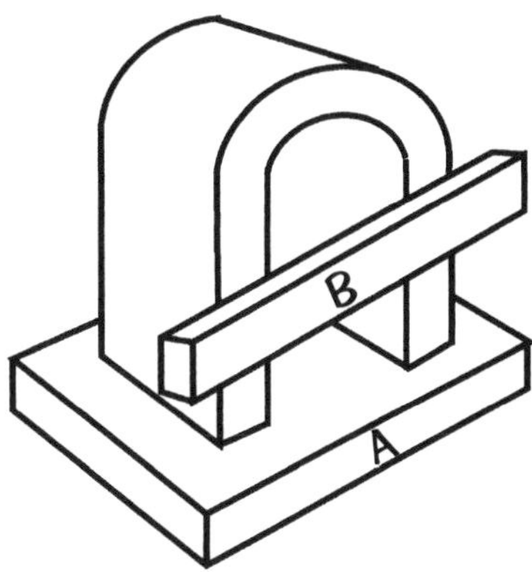

192. Magnetic Sphere

Magnets have at least two magnetic poles. If someone could build a magnet with just one pole, a great feat would be accomplished. Here's one modest proposal: Cut a steel sphere into irregular sections shaped as in the diagram. Magnetize all the sharp ends as south poles and all the other ends as north poles. Put the magnetized sections all back together again so they form a sphere. Will the magnetized sphere have a north pole on the outside? Has the south pole disappeared?

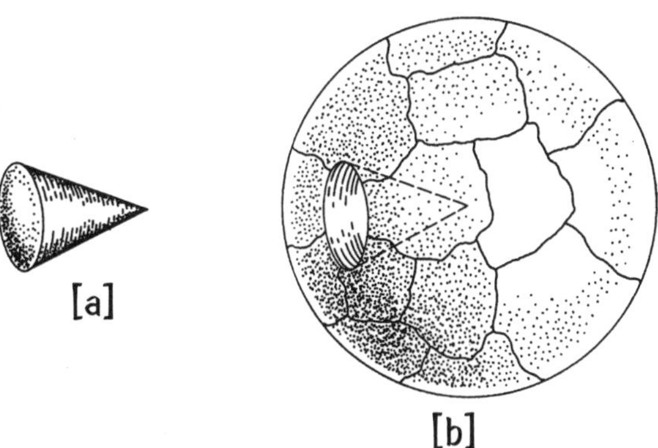

193. Two Compasses

Take two identical magnetic compasses, set one down on a table, and allow its needle to come to rest. Bring the second compass close to the first, shake it so its

The magnetic field around the chest, produced by the relatively large current in the heart, is about 10^{-6} gauss, and the smaller currents in the brain produce a magnetic field around the head of about 3×10^{-8} gauss. These values are much smaller than the earth's field (0.5 gauss) or even the fields associated with the currents in the electrical wires of a building (5×10^{-4} gauss).

Infinity is just time on an ego trip.

—Lily Tomlin

The opposite of a correct statement is a false statement. But the opposite of a profound truth may well be another profound truth.

—Niels Bohr

Faraday was once asked by a colleague, "Is Professor X always wrong?"

"He's not that consistent," replied Faraday.

—DANIEL KOSLAND JR.

A magnetic compass rarely points true north. While a compass in Chicago does align itself almost due north, one in Los Angeles tilts about 15° to the east, and one in New York leans almost 15° west of north.

—EUGENE HECHT

The actual universe is far more interesting than the universe of science fiction.

—LAWRENCE M. KRAUSS

needle is set into oscillation, and place it on the table close to one end of the needle of the first compass. What do you predict will happen to the compass needles?

194. Magnetic Work?

We know that magnetic fields never do any work on charged particles. Yet, a current-carrying wire placed in a magnetic field will gain kinetic energy as the wire accelerates in response to the magnetic force. How can that be?

195. Electric Shield

Will an electric shield that shields the inside of a volume from an external electric field also shield against an incident electromagnetic wave?

196. Wave Cancellation in Free Space

"When light waves cancel, where does the energy go?" is a question that assumes that the same energy is carried by the two interfering waves as is carried by the same two waves when they are separate. As such, this assumption is a misuse of the principle of superposition. Why?

197. Repulsion Coil I

A conducting metal ring around a vertical coil will levitate when a steady AC current passes through the coil (see page 75). If this metal ring around a repulsion coil is replaced by a stiff circle of nonconducting string, what do you predict will happen? Will there be an EMF around the circular string?

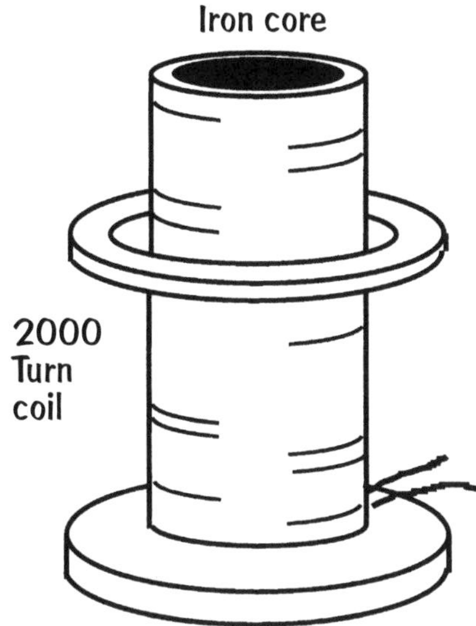

198. Repulsion Coil II

Why does a conducting metal ring around a repulsion coil jump when the AC current through the coil is quickly turned on?

199. Magnetic Tape

When a loop of magnetic audiotape picks up an electric charge, what shape do you predict the audiotape will take?

200. Kelvin Water Dropper

The Kelvin water dropper is an amazing device invented by Lord Kelvin that uses water to generate voltages of up to 15,000 volts. Cans *A* and *D* are connected electrically, as are cans *B* and *C*. Water drips through the two bottomless metal cans *A* and *B* and is collected in cans *C* and *D*. Almost immediately upon startup, the electrically neutral cans become electrically charged, one pair positively and the other pair nega-

Physics is not a religion. If it were, we'd have a much easier time raising money.

—Leon Lederman

The most powerful permanent magnets (samarium-cobalt or neodymium-iron-boron) product fields of about 0.3 tesla to 0.4 tesla.

James Clerk Maxwell died in 1879, the year Albert Einstein was born. The opening sentence of Einstein's first paper on relativity states: "Maxwell's electrodynamics . . . when applied to moving bodies, leads to asymmetries which do not appear to be inherent in the phenomena."

—P. Halevi

Knowledge advances funeral by funeral.

—Paul Samuelson

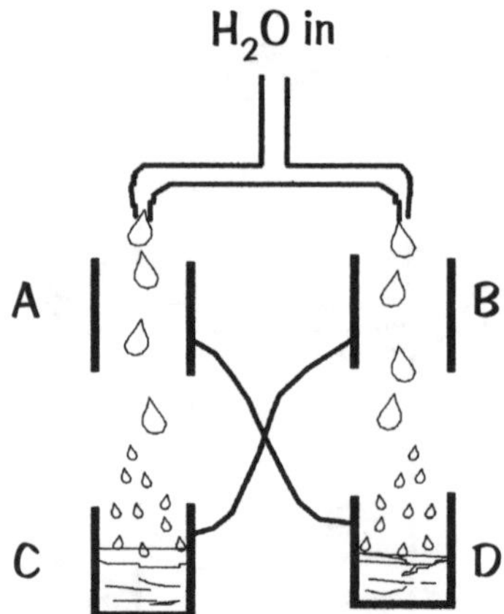

tively. The voltages developed can get so high that a small fluorescent lamp brought close to one of the cans will flash. How does this device work?

*201. Back EMF

Is the back EMF of a motor something that blocks energy flow to the motor? Explain.

Faraday read the papers Maxwell sent him with the bemusement of a tone-deaf man listening to Beethoven's quartets, understanding that they were beautiful without being able to appreciate just how. "I was almost frightened when I saw such mathematical force made to bear upon the subject, and then wondered to see that the subject stood it so well," Faraday wrote Maxwell.

—Timothy Ferris

*202. Axial Symmetry

A long positively charged wire along the axis of a long negatively charged metal tube creates an electrostatic situation that has axial symmetry. What happens to a neutral particle initially at rest between the two electrodes?

*203. A Ring Is a Ring . . . !

A uniform copper ring is threaded by a steadily increasing magnetic field so that the magnetic flux through the ring is changing at a constant rate. Use a voltmeter to measure the voltage across the metal ring. What do you predict?

*204. Electromagnetic Field Energy

Half the energy in an electromagnetic wave is in the electric field and the other half is in the magnetic field. Both the electric field and the magnetic field reach their maximum values simultaneously and their minimum values simultaneously. When the fields are simultaneously zero, where is the energy?

Physical concepts are free creations of the human mind, and are not, however it may seem, uniquely determined by the outside world.

—ALBERT EINSTEIN

*205. Levitating Top

A small spinning top made from a permanent magnet can maintain a stable levitating position in air above a slightly curved magnetic platform for several minutes. How does it work?

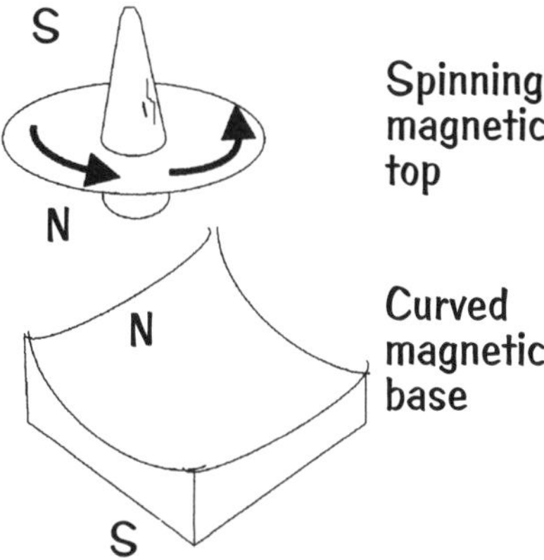

In my house there's this light switch that doesn't do anything. Every so often I would flick it on and off just to check. Yesterday, I got a call from a woman in Germany. She said, "Cut it out."

—STEVEN WRIGHT

*206. Levitating Mouse

Recently, a mouse has been levitated in a magnetic field. What is the physics here?

Science is the attempt to make the chaotic diversity of our sense-experience correspond to a logically uniform system of thought.

—ALBERT EINSTEIN

7 Bodies in Motion

In mechanics, Newton's laws are an excellent approximation to the behavior of nature, unless one is considering relativistic speeds. So let's find out how well you can apply them to the braintwisters, paradoxes, and fallacies we have gathered here. As one finds in most endeavors, a careful reading followed by a judicious choice of ideal physical properties and pertinent approximations should be considered first. If a clear solution does not reveal itself then, one would be wise to remove the idealizations one by one until a satisfactory explanation is achieved.

207. Superwoman

The person in the illustration is attempting to lift herself and the chair off the ground by pulling downward on the rope. The woman and the chair would move upward together. What do you predict will happen when she pulls on the rope?

208. Lifting Oneself by One's Bootstraps

Can the man in the illustration lift himself and the block off the ground? After all, he looks as though he

Archimedes, who was kinsman and a friend of King Hieron of Syracuse, wrote to him that with any given force it was possible to move any given weight, and emboldened, as we were told, by the strength of his demonstration, he declared that, if there were another world, and he could go to it, he could move this. Hieron was astonished, and begged him to put his proposition into execution, and show him some great weight moved by a slight force. Archimedes therefore fixed upon a three-master merchantman of the royal fleet, which had been dragged ashore by the great labors of many men, and after putting on board many passengers and the customary freight, he seated himself at a distance from her, and without any great effort, but quietly setting motion with his hand a system of compound pulleys, drew her towards him smoothly and evenly, as though she were gliding through the water.

—Plutarch

If it would take a cannon ball 3 1/3 seconds to travel four miles, and 3 3/8 seconds to travel the next four, and 3 5/8 to travel the next four, and if its rate of progress continued to diminish in the same ratio, how long would it take to go fifteen hundred million miles?

—ARITHMETICUS

I don't know.

—MARK TWAIN (FROM "ANSWERS TO CORREPSONDENTS")

A body in motion can maintain this motion only if it remains in contact with a mover.

—ARISTOTLE

is trying to lift himself up by his own bootstraps, which, Baron Munchausen's boastful stories notwithstanding, is impossible.

209. Springing into Action

A spring balance is hung from the ceiling by a long rope. A second rope is attached to the spring balance, pulled tight so the balance reads 100 pounds, and then anchored to the floor. If a 60-pound weight is now hung on the hook of the balance, what do you predict the balance will read?

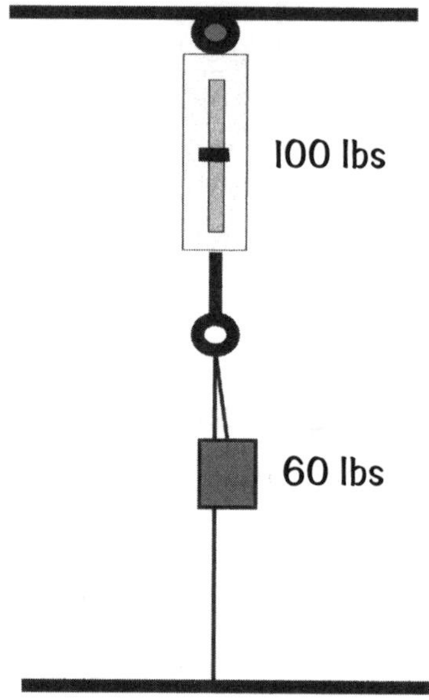

210. The Monkey and the Bananas

This problem is an old one and is said to have been invented by Charles Dodgson (also known as Lewis Carroll): A long rope passes over a pulley. A bunch of bananas is tied to one end of the rope, and a monkey of the same mass holds the other end. What will happen to the bananas if the monkey starts climbing the rope?

Assume the ideal rope and pulley: Neither has weight, the rope is extensionless, and there is no friction opposing the turning of the pulley.

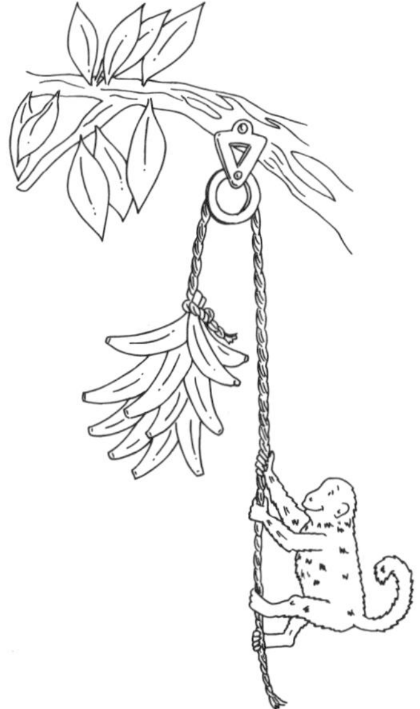

211. Hourglass on a Balance

An hourglass timer is being weighed on a sensitive balance, first when all the sand is in the lower chamber, and then after the timer is turned over and the sand is falling. Will the balance show the same weight in both cases?

212. How Much Do I Weigh, Anyway?

Even if you stand very still on an accurate scale, the reading keeps oscillating around your average weight. Why? As you *begin to step off the scale,* what do you predict for the immediate value of the scale reading?

213. A Bumpmobile

A woman stands on a wooden plank and hits one end of it with a heavy hammer. The plank and the woman

1 jelly doughnut (JD) = 10^6 joules
1 mosquito pushup = 1 erg = 10^{-13} JD

A new scientific truth does not triumph by convincing its opponents and making them see the light, but rather because its opponents eventually die, and a new generation grows up that is familiar with it.

—Max Planck

Newton and Huygens concluded that the centrifugal force associated with the Earth's rotation would cause it to bulge at the Equator and flatten out at the poles. Because the leading French astronomers and some theorists following the ideas of René Descartes came to the opposite conclusion, this problem was seen as a crucial test of the competing Newtonian and Cartesian systems of the world. The results of the French expeditions in the 1730s confirmed the predicted flattening at the poles and thus helped to ensure the victory of Newton over Descartes.

move together. You very likely did something of the sort as a child and found that you could propel yourself along the floor. Where is the external force? One can visualize the woman and the plank enclosed by a big box that still affords the woman enough space to swing the hammer so that the described motion ensues. The box then appears to jerk forward with no apparent outside help.

Doesn't this action violate Newton's first law? A body remains at rest or in a state of uniform velocity (constant speed in a straight line) unless acted upon by a net *external* force. The sliding friction between the plank and the floor is a relevant horizontal external force. Unfortunately, this sliding frictional force acts to oppose the motion of the plank, so how can friction propel the plank forward?

214. The Wobbly Horse

There is an old toy horse that has straight legs that swing forward and backward only at the connections to the sides of the horse's body. When pulled forward on a tabletop by a string, the horse wobbles forward. Imagine that the toy horse is arranged as shown in the diagram. The horse begins about a foot from the edge of the table and is pulled forward by the constant applied force along the thread that passes over the edge of the table to support the hanging object (here consisting of several paper clips). What do you predict for the behavior of the horse after it begins to move forward?

215. Two Cannons

What will happen if two identical cannons are aimed at each other and the shells fired simultaneously and at the same speeds? One cannon is higher than the other, but the two are perfectly aligned.

216. The Law of Universal Gravitation

Newton's law of universal gravitation is sometimes expressed by the equation $F = GMm/d^2$, where F represents the force between two objects of masses M and m, d is the distance between their centers of mass, and G is the gravitational constant. Is this equation a correct formulation of Newton's law of universal gravitation?

For an interesting application, consider a carpenter's square with its center of mass at the point C in the

I do not know what I may appear to the world, but to myself I seem to have been only like a boy playing on the sea shore, and diverting myself in now and then finding a smoother pebble or a prettier shell than ordinary, while the great ocean of truth lay all undiscovered before me.

—Isaac Newton

We should regard the present state of the universe as the effect of its antecedent state and the cause of its subsequent state. An intelligence acquainted with all the forces of nature, and with the positions at any given instant of all the parts thereof, would include in one and the same formula the movements of the largest bodies and those of the lightest atoms.

—PIERRE-SIMON DE LAPLACE

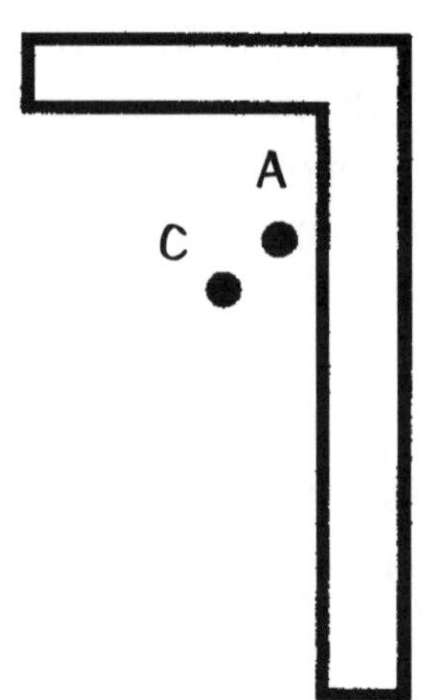

space between its two arms. A small spherical body placed at *C* should produce an infinite attraction because the distance between the centers of mass is zero! This result is clearly nonsensical. Indeed, one may even place the small sphere closer to the inside corner of the square at *A* to produce a movement that looks like a repulsion! How does one resolve this dilemma?

217. Balancing a Broom

A meter stick will balance on your finger if you support it at its center of gravity—the midpoint. The two halves have equal weights. A broom will also balance on your finger if you support it at its center of gravity. Suppose you cut the broom into two parts through the center of gravity and weighed each part on a scale. Would their weights be equal?

218. *Vive la Différence*

Is there a significant difference in the position of the center of mass of a man and a woman? The following demonstration, sometimes used as a "party trick," can reveal some information. A kneeling woman first places her elbows, arms, and hands together (as if "praying"), with the elbows touching the knees and the forearms along the floor. A matchbox or a similar object is placed at her fingertips. She then clasps her hands behind her back and is instructed to knock the matchbox over with her nose without tipping over herself. In general, women can perform this task, whereas most men cannot. Why not?

Colleges are places where pebbles are polished and diamonds are dimmed.

—ROBERT G. INGERSOLL

219. Balance Paradox

The two equal-weight objects shown are free to slide on horizontal bars attached to a sort of pantograph. The pantograph is constructed so that the vertical links always remain vertical and the longer horizontal bars always remain parallel as the system tilts one way or the other. The left-hand object has just been moved out farther than the right-hand object. Which end, if any, will go down?

[T]hat one body may act on another through a vacuum, without the mediation of anything else, by and through which their action and force may be conveyed from one to another, is to be so great an absurdity, that I believe no man who has in philosophical matters a competent faculty of thinking can ever fall into it.

—ISAAC NEWTON

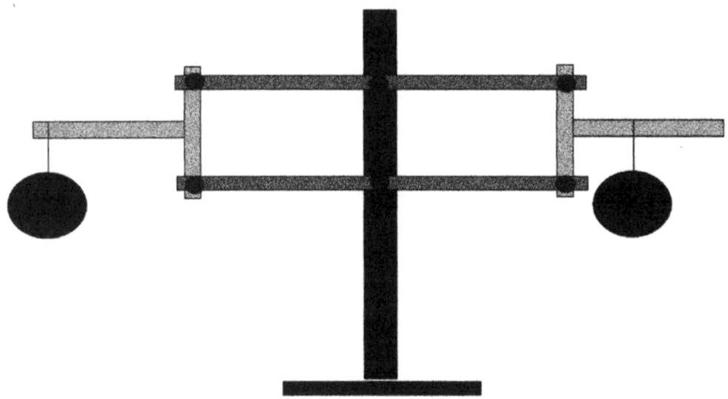

220. Tightrope Walker

Walkers along high wires carry a heavy horizontal bar. You would think that this extra weight makes each step harder to achieve than with a lighter bar. What's going on here? And how would a physicist distribute the bar's weight?

At Cambridge, Newton was the very model of an absent-minded professor. His amanuensis, Humphrey Newton (no relative), wrote that he never knew Newton "to take any recreation or pastime either in riding out to take the air, walking, bowling, or any other exercise whatever, thinking all hours lost that was not spent in his studies." He often worked until two or three o'clock in the morning, ate sparingly and sometimes forgot to eat altogether. When reminded that he had not eaten, he would go to the table and "eat a bite or two standing." Newton rarely dined in the college hall; when he did, he was apt to appear "with shoes down at heels, stockings untied, surplice on, and his head scarcely combed." It was said that he often delivered his lectures to an empty hall, apparently with as much satisfaction as if the room had been full of students.

—I. Bernard Cohen

221. Balancing an Upright Stick

Generally speaking, bodies with low centers of gravity are more stable than those with high. For example, a stub of a pencil can be stood on its flat end very easily, but it is much harder to stand a long stick on its flat end. Paradoxically, however, a long stick with its higher center of gravity is much easier to balance on the tip of a finger than a short pencil. Why?

222. Racing Rods

Consider the contraption shown: Rods *A* and *B* are identical in length and mass, except that rod *B* has a small spherical ball attached at the end. The two rods are free to rotate about the rigid, frictionless horizontal pivot. Suppose the rods are released simultaneously from rest when they are standing straight up from the pivot. What do you predict about their elapsed times to reach the lowest position?

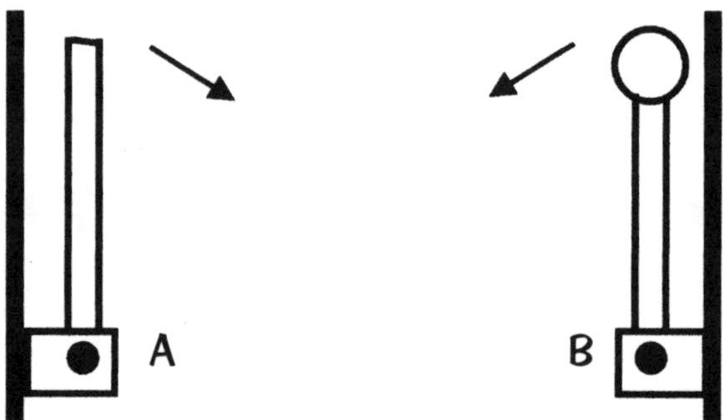

223. Magic Fingers

Support a uniform stick (such as a meterstick or a long dowel) with both index fingers so the stick is not horizontal but shows a definite slant. Start with the support fingers nearly equidistant from the center of gravity.

Before moving anything, predict which side will move first: the higher or the lower index finger? Now move the index fingers together slowly and observe their initial motion. What is the explanation?

> Science has been seriously retarded by the study of what is not worth knowing and of what is not knowable.
>
> —GOETHE

224. The Soup Can Race

If you simultaneously release a solid sphere, a solid cylinder, and a hoop at the top of an incline, every time the sphere will win the race. The uniform-density sphere wins all these races no matter how its mass and/or radius compare with the masses and radii of the hoops and the cylinders.

A common variation on this race is to compare the descent down the incline between a can of chicken noodle soup and a can of something like cream of broccoli soup. What do you predict? Does the result depend on the can dimensions? On the masses? What does the acceleration down the incline depend on?

225. The Tippe Top

The plastic tippe top is shaped like a mushroom. If you release the spinning toy on the floor, it soon inverts itself while continuing to spin. If the top is spinning clockwise when viewed from above before inversion, what direction is the tippe top spinning after inversion occurs? What role does friction play in the inversion?

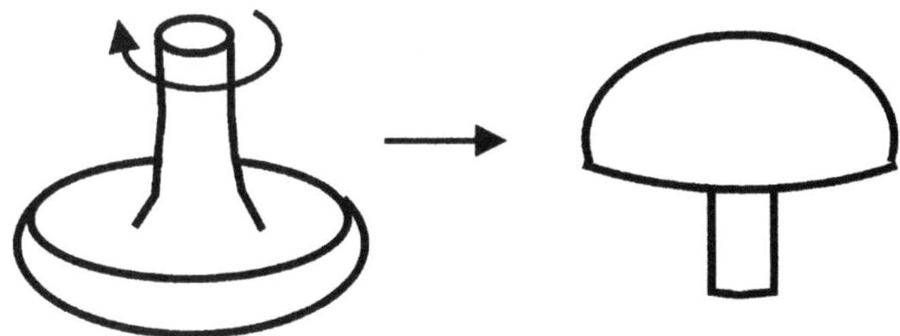

Now I a fourfold vision
see,
And a fourfold vision is
given to me;
'Tis fourfold in my
supreme delight
And threefold in soft
Beulah's night
And twofold Always.
May God us keep
From Single vision &
Newton's sleep!

—WILLIAM BLAKE (REBEL AGAINST THE MECHANICAL SINGLE VISION, OR LINEAR THINKING, OF NEWTON)

226. The Mysterious "Rattleback" Stone

If you spin a "rattleback" stone (a long stone with a rounded bottom) in the "wrong" direction, it will quickly stop, rattle up and down for a few seconds, and then spin in the opposite direction. Most have an ellipsoidal bottom, with the long axis of the ellipsoid aligned at an angle of five to ten degrees to the long axis of the flat top. What is the physics behind this mysterious behavior?

227. The Case of the Mysterious Bullet

Two ideal bullets, identical in shape, size, and mass, strike the same target with the same speed just before the collision. Force meters at the target register two times the force value for bullet *A* compared to bullet *B*. Is one of the force meters faulty?

228. Centers of Mass of a Triangle and a Cone

The center of mass *CM* of an isosceles triangle is located at a point one-third up the altitude of the

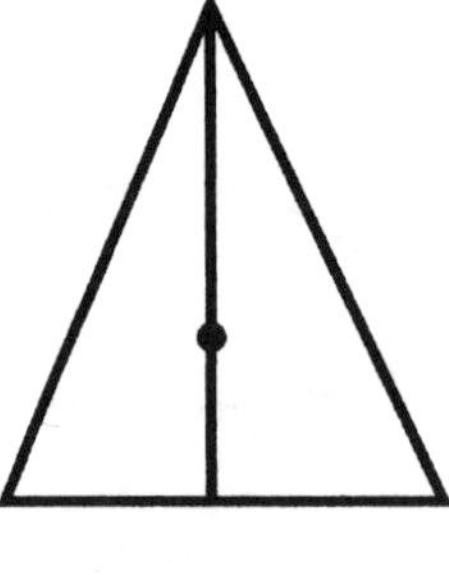

a

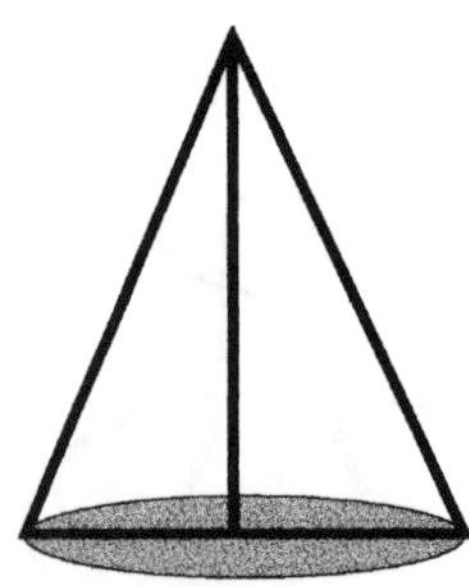

b

triangle above the base (see part a of the diagram). Now consider a right circular cone of the same cross section (see part b of the diagram). Is its center of mass also located at a point one-third up the altitude of the cone?

229. Staying on Top

If you shake a bucket partially filled with apples of various sizes for a couple of minutes, the biggest apples tend to end up on top. Why?

230. Antigravity

Try this demonstration at home: Hook two paper clips together and insert the opposite ends into two straws. Place the straws together over a third straw or pencil, balance a marble at the lower end, and gradually separate the straws at the upper end. Surprisingly, the marble will appear to roll uphill! How can the marble appear to defy gravity?

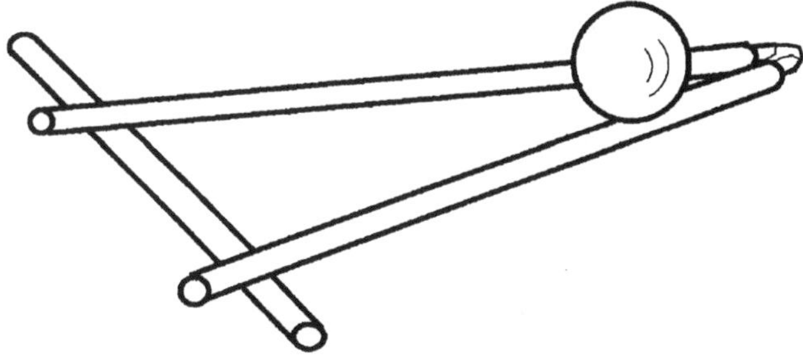

231. Which Path?

Imagine four inclined planes set up to form a rhombus, as shown on page 91. Two identical balls are released simultaneously from A so that one rolls along ABC and the other along ADC. What do you predict?

Postulate 1: Knowledge is power.
Postulate 2: Time is money.
But power = work/time. Using postulates 1 and 2, this can be written as knowledge = work/money. Solving for money, we get money = work/knowledge—that is, as knowledge approaches zero, money approaches infinity, regardless of work done. Conclusion: The less you know, the more you make.

The visiting speaker was at the chalkboard deriving an algebraic relation and was embarrassed when the result was negative instead of positive. "I must have made a mistake," he said. From the audience, Pauli (or was it Dirac?) corrected him, "More likely, an odd number of mistakes."

Newton's writings on the subject of alchemy amounted to about 650,000 words. It could be said, therefore, that his work in physics was something of an interruption of this larger lifelong quest. In this context, Morris Berman writes: "The centerpiece of the Newtonian system, gravitational attraction, was in fact the Hermetic principle of sympathetic forces, which Newton saw as a creative principle, a source of divine energy in the universe. Although he presented this idea in mechanical terms, his *unpublished* writings reveal his commitment to the cornerstone of all occult systems: the notion that mind exists in matter and can control it (original participation)."

—MORRIS BERMAN

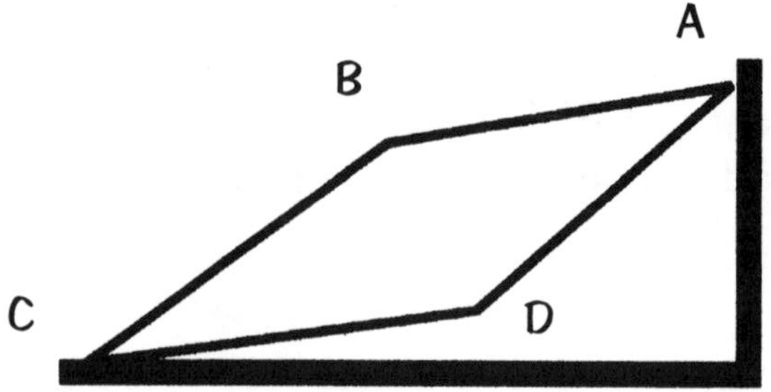

232. Is the Shortest Path the Quickest Way Home?

Imagine a frictionless bead released at point *P* in the diagram. Sketch the shapes of a piece of wire giving the fastest (i.e., shortest time) paths from point *P* to points *A*, *B*, and *C*. Straight lines just don't cut it! You will need paths that start out with a steep vertical drop. Why?

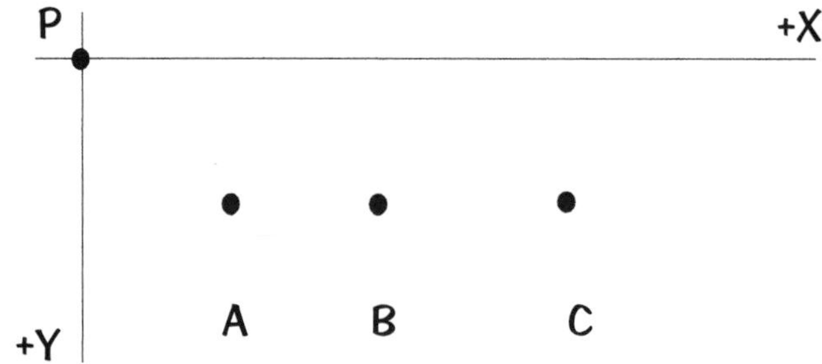

233. The Unrestrained Brachistochrone

Two identical marbles start with the same initial speed and roll on two different tracks, marble *A* rolling on a straight track having a slight slope from start to finish and marble *B* rolling up and down the valleys and hills of the second track from the same start position to the same finish position, all the while maintaining contact with its track. If both start simultaneously from rest, which do you predict will finish first? What is the physics?

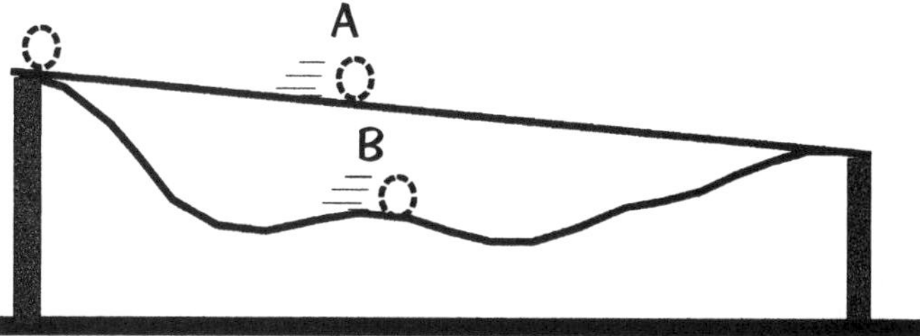

Isaac Newton spent years trying to date the world's end. He leaned toward 1948.

*234. Tilting Rods

Take two identical uniform cross-section rods, leave one bare, and attach a heavy object to the upper end of the other. Place them with their lower ends against a wall but at the same angle from the vertical. Now release the rods simultaneously so that they begin to fall to the floor. What do you predict? Suppose that the heavy object were fastened lower down on the rod. What happens now?

Part of thinking is its cruelty, aside from its contents. It is the process of detachment from everything else, the ripping, the wrenching, the sharpness of cutting.

—Elias Canetti

*235. Faster than Free Fall

Can objects fall with an acceleration greater than the acceleration of gravity? For an answer, try the following: Make a shallow cup by cutting about 3 centimeters off the bottom of a Styrofoam cup. Push one side in so it points about 30 degrees away from the vertical. Now

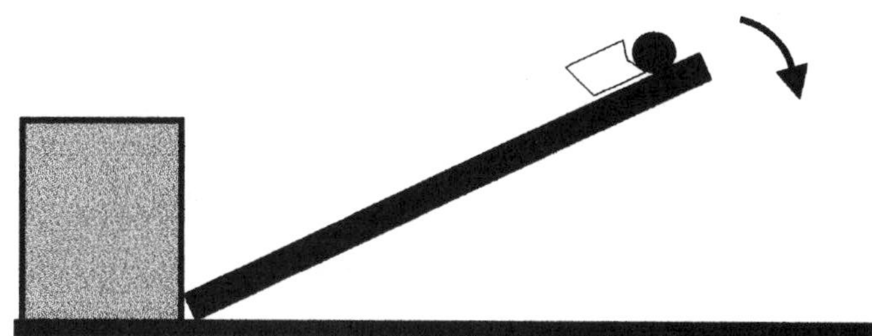

It was absolutely marvelous working for [Wolfgang] Pauli. You could ask him anything. There was no worry that he would think a particular question was stupid, since he thought all questions were stupid.

—Victor Weisskopf

tape the cup to a long yardstick (or a meterstick) so its center is about 6 centimeters from the end and the dented side is facing the near end of the yardstick.

Hold the yardstick at thirty to forty degrees to the horizontal, with one end on the floor against a solid object such as the wall. Place a marble against the cup so it rests against the dent in the cup. Notice that the lip of the cup is higher than the ball. Now release the ruler. *Voilà!* The marble will drop into the cup, indicating that the acceleration of the cup was greater than the acceleration of gravity ($g = 9.8$ m/sec^2) and hence drops away from the marble. How can this result be true?

*236. Racing Cylinders

You have two cylinders of identical size and mass. They are made of two materials of different densities. The cylinder made of the higher-density material is hollow. How can you tell which one is which?

*237. Friction Helping Motion

We usually think of friction as opposing motion. Can friction sometimes help motion? Yes. In fact, we experience the beneficial effects of friction numerous times every day. When a car accelerates forward from a stop sign, the static frictional force between the road and the drive wheels produces this forward acceleration.

Now to the case at hand: Can the friction produce a *greater* acceleration of an object? Consider the test case of wrapping a rope several times clockwise around a horizontal cylinder (a large spool, say) with mass M and pulling on it with a horizontal force F to the right so the cylinder rolls without slipping. Can you show that the horizontal acceleration of the cylinder is 4/3 F/M—that is, *greater than* the acceleration F/M produced when the cylinder is simply pulled along the sur-

Nec fas est propius mortali attingere divos (Nearer the gods no mortal may approach).

—Edmund Halley (on Newton)

Nature, to be commanded, must be obeyed.

—Sir Francis Bacon

[Testifying before a congressional committee, Robert R. Wilson] launched into an impassioned argument for the [Fermilab] accelerator, and Senator Pastore was being impassioned right back. Among other things, the senator said: "Is there anything connected [with] this accelerator that in any way involved the security of the country?" Wilson responded: "No, sir; I do not believe so." Senator Pastore persisted: "Nothing at all?" Finally Wilson said: "[The accelerator] has to do with: Are we good painters, good sculptors, great poets? . . . It has nothing to do with defending our country, *except to make it worth defending.*"

—Dennis Flanagan

face without any rotation or friction? Remember, the applied force *F* and the static frictional force are the only external horizontal forces acting on the cylinder. The conclusion is inescapable: The static frictional force must be in the same direction as *F*, helping the cylinder move faster. Is there anything wrong with this conclusion?

An angry human female develops an acceleration of about 100 g's at the foot when stalking off in high heels.

*238. Obedient Spool

Take a spool, preferably a large one like the kind that wire cable comes wound on. Wrap a wide ribbon around the axle of the spool so it "peels off" the bottom. Now try pulling on the ribbon. The results will be contrary to expectations. By increasing the angle between the ribbon and the vertical, the spool can be made to roll toward you. By decreasing this angle, the spool can be made to roll away from you. An intermediate value of the angle can be found where the spool will simply skid without rolling along the floor. How can one explain this strange behavior?

Colleges hate geniuses, just as convents hate saints.

—Emerson

*239. "And the Winner Is . . ."

In a race down an incline, a solid, uniform-density sphere will always beat a solid cylinder. The latter will always beat a hoop. What do you think will happen when a solid cone is rolled *straight down* the incline in a race against the other three? How can one make the solid cone roll straight?

In 1912, when the Nobel Prize selection committee announced that Nikola Tesla and Thomas Edison had been chosen to share the award in physics, Tesla indignantly turned it down. He considered himself to be a discoverer of principles and Edison a mere inventor of appliances and refused to be linked with Edison. The 1912 Nobel Prize in physics went to Nils Gustav Dalin, inventor of automatic regulators for gas-burning lights in lighthouses and buoys.

*240. Getting into the Swing of Things

Most children are able to start a swing from rest without any outside help and without touching the ground or other objects. How is this feat possible?

NEWTON AND THE MOON I

In order to calculate the rate of the moon's fall to Earth, the exact distance from the surface to the center of the Earth is required, and this in turn depended on the size of a degree of latitude. As he was away from Cambridge and his books [in 1666], Newton used a current estimate of 60 miles to one degree of latitude, which was about one-sixth too little. As a result Newton would have calculated that the rate of the moon's fall to Earth was such that it would fall 12 feet during the first hour, whereas if the inverse square law held true the figure should have been 16 feet in the first hour. Newton could not be satisfied with such a discrepancy. He concluded with characteristic caution that some other factor, perhaps the Cartesian vortices, was at work and put the problem aside for nearly twenty years.

—VINCENT CRONIN

*241. Pumping on a Swing in the Standing Position

If a child standing on a swing is given a slight push, he can soon learn, by trial and error, to gain height by a pumping process that amplifies the initial deflection. What is the physics behind this process?

*242. Pumping on a Swing in the Sitting Position

Do you pump differently when you are sitting on a swing from the way you pump in the standing position?

*243. Spinning Wheel

A man is holding a bicycle wheel with a lead-filled tire in front of his chest, with one end of the horizontal axle in each of his outstretched hands. The vertical wheel is set spinning between his arms. Suppose the man wishes to rotate the plane of the spinning wheel slightly to the left about its present vertical axis—that is, have the

axle remain horizontal while its left end moves closer to his ribs and its right end moves farther away. Will pushing forward with his right hand and pulling backward with his left do the trick?

> Newton was not only the greatest genius that ever existed, but also the most fortunate, inasmuch as there is but one universe, and it can therefore happen to but one man in the world's history to be the interpreter of its laws.
>
> —Pierre-Simon de Laplace

*244. Collision with a Massive Wall

Imagine a ball of mass m moving with a speed v colliding head-on with a massive wall. If the collision is elastic, the ball simply recoils with the same speed v. But if this is true, then the kinetic energy of the ball, $1/2\ mv^2$, is conserved, but its momentum, mv, is not, because its velocity (a vector) is now directed in the opposite direction.

The thoughtful reader may say that the laws of the conservation of momentum and energy should be applied to the total system consisting of the ball and the wall (or wall + earth). Correct. Then the momentum change of the ball (final momentum minus initial), $m(-v) - mv = -2mv$, plus the momentum change of the wall + earth, $MV - M(0) = MV$, must sum to zero. But then the energy will not be conserved, for the total energy before the collision is $mv^2/2$, and the total energy after the collision is $mv^2/2 + MV^2/2$. What's the way out of this paradox?

> Science can purify religion from error and superstition, while religion can purify science from idolatry and false absolutes.
>
> —Pope John Paul II

*245. Executive Toy: Newton's Cradle

There is a popular toy consisting of five steel balls, all of the same size and mass, hanging side by side in contact in a row on bifilar supports. Pull one back at one end and release it, and one ball moves off the other end, etc., etc. If two are pulled back at the same end and dropped together, two move out together at the other

Newton was not the first of the age of reason. He was the last of the magicians. . . . He looked on the whole universe and all that is in it *as a riddle,* as a secret which could be read by applying pure thought to certain evidence, certain mystic clues which God had laid about the world to allow a sort of philosopher's treasure hunt to the esoteric brotherhood. He believed that these clues were to be found partly in the evidence of the heavens and in the constitution of elements (and this is what gives the false suggestion of his being an experimental natural philosopher), but also partly in certain papers and traditions handed down by the brethren in an unbroken chain back to the original cryptic revelation in Babylonia. He regarded the universe as a cryptogram set by the Almighty.

—JOHN MAYNARD KEYNES

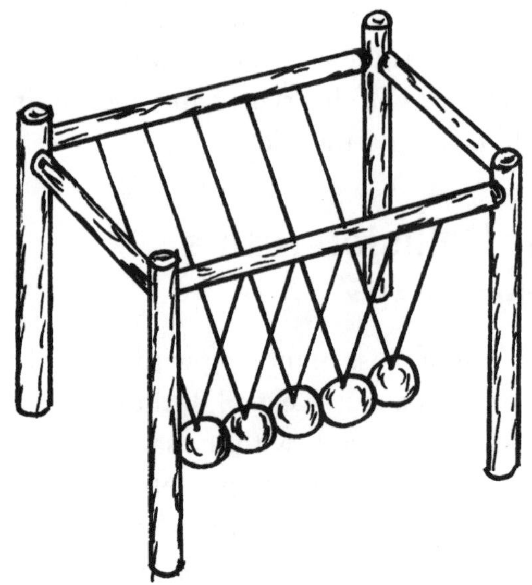

end. Pull back three, release together, and three move out at the other end. Pull back four, and . . . Pull back five, and . . . The balls can obviously count! How do they know?

*246. Hammering Away

Why is it easier to drive a small stake into the ground with a heavy hammer (even swung gently) than with a light one, although the latter can be swung with great speed, thus giving it enormous energy? Contrast this situation with forging: here the hammer is much lighter than the anvil. Why?

*247. Velocity Amplification

When a small ball is placed on top of a large ball and the two are dropped together, something dramatic happens when the combination rebounds from the floor. The small ball will take off and can reach a height almost nine times its original height! Any idea why?

*248. Superball Bounce

A superball approaches the floor with a forward horizontal velocity and unknown spin. After bouncing from the floor, its velocity is still forward and its spin is zero. What was the direction of the initial spin?

*249. Ring Pendulum

A uniform hoop is supported so it hangs in the vertical plane by a knife edge. Set into oscillation in the plane, this physical pendulum has a natural period of oscillation. Subsequently, symmetric sections starting from the bottom of the hoop are cut off. What is the period of the half hoop? Of the quarter hoop? What is the surprise in the behavior of this oscillating system?

*250. One Strange Pendulum

A simple pendulum swings freely at its natural frequency f_0 when suddenly its point of support begins to oscillate up and down in simple harmonic motion at frequency f. At what frequency must f be to rapidly increase the amplitude of each swing of the pendulum?

If you shut your door to all errors, truth will be shut out.

—Rabindranath Tagore

Newton and the Moon II

The appearance of a comet in 1680, and of a second comet in 1682 that moved in a direction opposite to the planets, aroused new interest in the paths of these conspicuous but short-lived phenomena, and turned Newton's mind back to astronomy. In June 1682, at a meeting of the Royal Society, Newton heard about the work of Monsieur Jean Picard, who was mapping France with sophisticated instruments, and had found the length of a degree to be 69.1 miles. Newton repeated his former calculations using Picard's value and found that the rate of the moon's fall to Earth exactly corresponded to the inverse square law.

—Vincent Cronin

8 Stairway to Heaven

STRUCTURES FORM THE BASIS OF ALL MATErial objects, from the structure of the atom to the structures inside living organisms to the human-made structures on earth all the way up to the structures of the universe. In this chapter we concentrate on familiar structures and some of their basic foundations, as well as some of their peculiarities. You will encounter jumping fleas, I beams, and bursting pork sausage amid these challenges.

251. The I Beam

Steel girders used in construction often have the cross-sectional shape of an I beam, with most of the material collected in large flanges at the top and the bottom, and with the joining web rather thin. Why is this particular shape so universal?

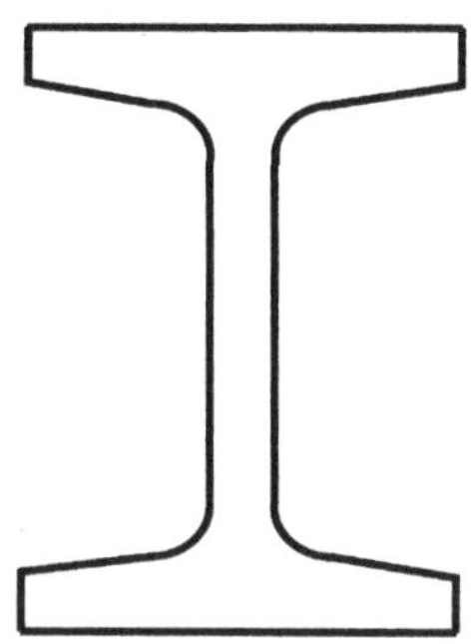

252. The Aluminum Tube

A solid aluminum rod and an aluminum tube of the same diameter made from the same material are not the same strength when identical bending forces are applied. What do you predict? Why?

253. Two Pulleys

Two identical pulleys with their centers at the same level are connected by a belt. The pulley on the left is the drive pulley. Is the maximum power that can be transmitted by the belt greater when the pulleys rotate clockwise or counterclockwise?

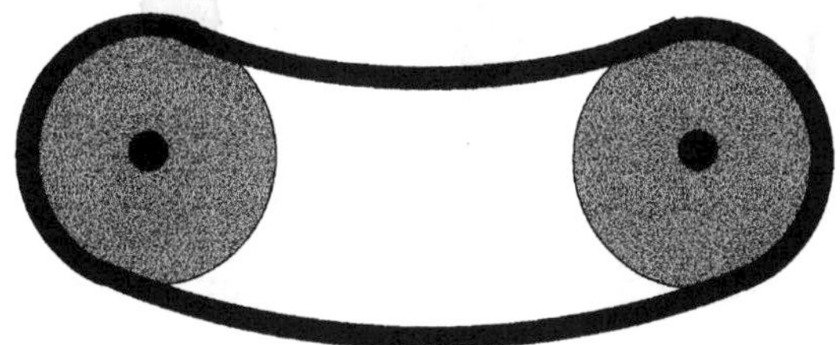

As far as the laws of mathematics refer to reality, they are not certain; as far as they are certain, they do not refer to reality.

—Albert Einstein

Write the following sequence of numbers: 113355. Cut it in half, and divide the second half by the first. The ratio 355/113 is an excellent approximation to *pi.*

A huge superball, made of a polybutadiene-based material known as Zectron, was created for a special event in Melbourne, Australia. Dropped from the twenty-third floor of a hotel, the ball rebounded to the fifteenth story, plummeted again, and smashed a parked convertible.

254. The Tensegrity Structure

The structure shown here is a tensegrity tower—a tower built with rods under compression and wire under tension only. None of the solid rods touches other rods, but wires connect the appropriate rod ends. How does it support itself?

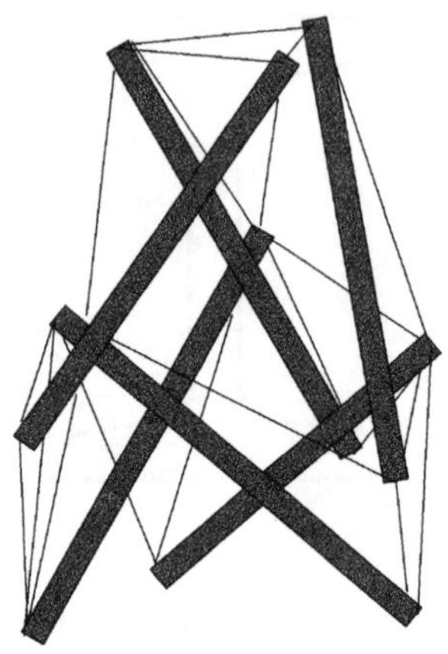

This fool [Copernicus] wishes to reverse the entire science of astronomy; but sacred scripture tells us that Joshua commanded the sun to stand still, and not the earth.

—Martin Luther

255. Vertical Crush

Most bricks used in buildings have a density of about 120 pounds per cubic foot (1926 N/m^3) and a crushing strength of at least 6,000 pounds per square inch (40 MN/m^2). As a result of the high crushing strength, a brick tower or building that is 7,000 feet tall (~ 2 km) could be built that would sustain the load! However, in fact, most brick buildings are much shorter and rarely support loads that are more than 3 percent of the crushing weight. Yet some of these short buildings have fallen over, even with so great a safety factor. What usually happens?

256. The Boat on High!

In many parts of the world one sees canals that are suspended as a bridge over low areas. Does the net load on the bridge change when a boat passes over the bridge portion of the canal?

257. Double the Trouble?

Two pieces of string are identical except one piece is twice the length of the other. Each is tied to a rigid support at one end, stretched taut, and subjected to the same sudden jerk at the unattached end. What do you predict will happen?

258. Boat's Anchor

A boat's anchor can be a massive hook tethered to the boat by a sturdy metal chain.

Yet, anchor chains do get broken, especially by inexperienced sailors. What usually happens?

259. Two Bolts

The illustration on page 106 shows two identical bolts held together, with their threads in mesh. While holding bolt *A* stationary, you swing bolt *B* around it. Don't let the bolts turn in your fingers. Will the bolt heads get nearer, move farther apart, or remain at the same distance?

Engineering is like a black hole: it makes you feel dense as it sucks you in.

—Curt Hepting

[An arch is] two weaknesses which together make a strength.

—Leonardo da Vinci

How to tell how tall your child will be: For girls, take the father's height and subtract five inches. To that number add the mother's height and then divide the total by two. For boys, take the mother's height and add five inches. To that number add the father's height and then divide the total by two. The formulas work best when the mother and the father are both either short, tall, or average in height for their sex.

All movement tends to maintenance, or rather all moved bodies continue to move as long as the impression of the force of their motors (original impetus) remains in them.

—LEONARDO DA VINCI

Kepler published all his laws [of planetary motion] by 1618, three hundred years after Giotto [a painter] intuited that the key to the accurate rendering of nature was the conic section.

—LEONARD SHLAIN

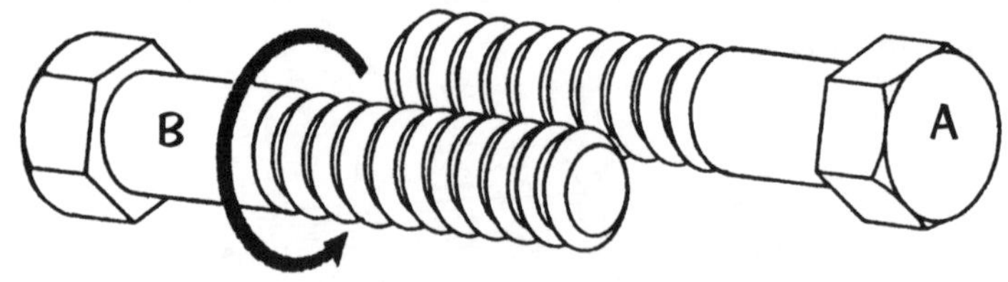

260. Tree Branching

A tree must transport nutrients between its central trunk and outermost leaves along a reasonably direct path. Why then can't a tree sustain each of its leaves with a separate branch? That is, why is the branching pattern in diagram (a) so much more common in nature than the explosive pattern in diagram (b)?

a

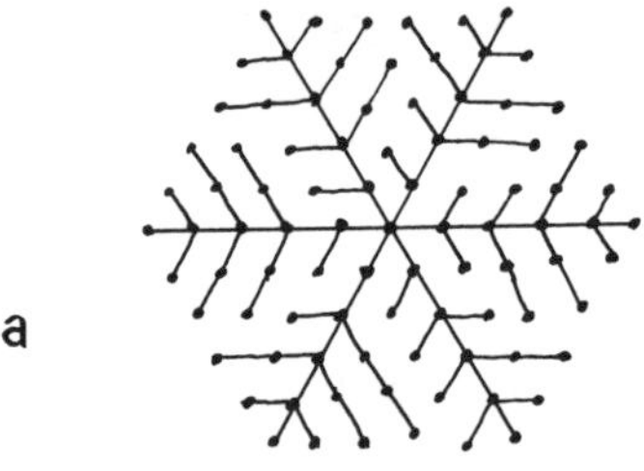

b

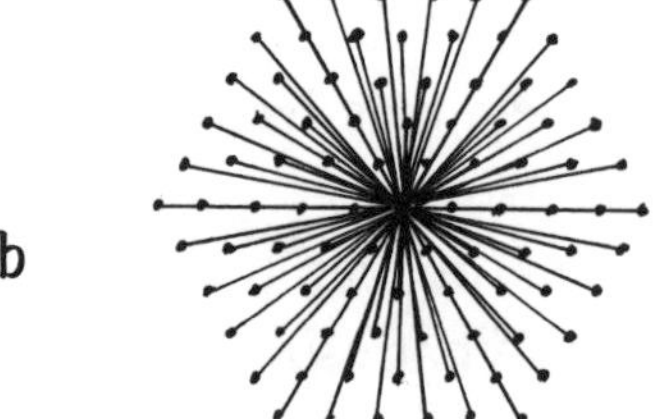

261. Hurricane Winds

The force exerted on a house by a 120-mile-per-hour hurricane wind is about twice the force exerted by a 60-mile-per-hour gale, right?

262. The Structural Engineer

At a New Year's Eve party, a storm was raging outside when a structural engineer was overheard saying: "This house was designed for stiffness, not for strength!" Should the owner be concerned?

A spider's dragline silk is, ounce for ounce, five times as strong as steel and five times as impact-resistant as bulletproof Kevlar.

263. My Arteries Are Stiff!

We have all heard people complain that their joints are stiff, especially the day after a strenuous workout. But what we cannot sense is the stiffness of our arteries all the time in all age groups, for the arteries are much stiffer than most other biomaterials in the human body. Why should the artery walls be so stiff?

By stiff, we can refer to some examples. Most engineering materials stretch less than 1 percent of their length, and most construction metals stretch less than 0.1 percent of their length. These are known as stiff materials. In contrast, many biomaterials can stretch 50 percent to 100 percent, such as the urinary bladder membrane in a young person.

I do not feel obliged to believe that the same God who has endowed us with sense, reason, and intellect has intended us to forgo their use.

—GALILEO

264. The Archery Bow

The instructions that accompany an archery bow usually remind the archer not to snap the bow without an arrow inserted to be propelled. Why not?

No one else I ever knew could copy a dozen numbers down wrongly, add them up wrongly, and come up with the right answer.

—LORD BOWDEN'S REMARK ABOUT LORD RUTHERFORD

265. The Pork Sausage Mystery

The skin of a pork sausage can burst during frying if the pressure inside grows large enough. In which direction is the skin more likely to break—longitudinally down the length of the sausage, or circumferentially around the body?

The intention of the Holy Ghost is to teach us how one goes to heaven, not how heaven goes.

—Galileo (quoting Cardinal Baronius)

Nowhere in [Newton's] *Principia* is there even a sketch of a valid proof that a body subject to an inverse-square force must move in a conic-section orbit.

—Robert Weinstock

The significant problems we face cannot be solved at the same level of thinking we were at when we created them.

—Albert Einstein

266. My Car Is a Steel Box!

Practically all cars today have a body shell that is essentially a type of steel box. Decades ago, the body was made differently, because panels were bolted onto a frame that had been bolted to an X-style or H-style chassis. Ignoring cost differences, why the deliberate shift to the steel box manufacture of modern cars?

267. Balloon Structure

Some domed stadiums have a bubblelike roof made of fabric that must be kept up by air pressure. In cold or rainy climates one also sees tennis courts and swimming pools covered by similar bubblelike fabric roofs. How can just a few fans keep these bubbles inflated properly?

*268. The Open Truss

Why is the open 3-D truss, composed of tetrahedra, so strong for its weight?

*269. Jumping Fleas

Fleas can jump to a height of 33 cm—more than 1 foot, or a hundred times their own length!—developing an acceleration of 140 g's. If a human could do that well in proportion to body height, the person could jump over a 50-story building. Why can't we?

*270. The Scaling of Animals

When an animal is scaled upward or downward in size, the weight increases as the cube of the linear dimensions—so that simultaneously doubling the height, the length, and the width of an animal will make its weight

eight times greater. The bone strength and the muscle strength increase as the cross-section—that is, the square of the linear dimension. Therefore, a "doubled" animal would have bones and muscles only four times as strong to carry eight times the weight.

But nature is very clever and does not design inadequate animals unable to support themselves! How big should the leg bones be to support eight times the original weight? And what should happen to the ribs and to the vertebrae?

*271. A Staircase to Infinity

Bricks are stacked so that each brick projects over the brick below without falling. Can the top brick project more than its length beyond the end of the bottom brick?

*272. Cowboy Lasso

How does a cowhand keep a rope loop spinning? Is there a minimum rotational speed?

Galileo died on January 8, 1642 in Italy. It was then 1641 in England. Newton was born on December 25, 1642 in England. It was then 1643 in Italy—the difference being the different times of adoption of the Gregorian calendar reform, plus different starting dates of the year. On a uniform calendar, Galileo died January 8, 1642 and Newton was born on January 4, 1643.

—Stanley E. Babb Jr.

Mark Twain would often say, "I was born with the comet, and I'm going to go out with the comet." And he did! He was born in 1835, during an appearance of Halley's comet, and died in 1910, during its next appearance.

9 Life in the Fast Lane

TRANSPORTATION OFFERS US THE OPPORTUNITY to combine the concepts from the two previous chapters on mechanics and structures into challenges that promise to enhance one's understanding of human-made machines. Baby carriages, bicycles, automobiles, and many other modes of transportation will be examined in this chapter. Static friction, kinetic friction, and rolling friction play significantly different roles in many of these challenges, so be careful. After all, the vehicles know the difference!

273. The Baby Carriage

Is a baby carriage with 2-foot wheels easier to push than one with 1-foot wheels?

274. The Falling Cyclist

Imagine you are riding a bicycle along a straight path when, suddenly, due to unevenness in the road or a wind gust, you find yourself tilting to one side. A beginner will instinctively try to steer to the other side, and soon he will have bruises to show for this response. In contrast, an experienced cyclist steers into the direction of the fall. Why?

275. Sudden Stops

On cars, the front brakes either are bigger drum brakes than the rear brakes or are disk brakes. Disk brakes do not overheat as easily as drum brakes because disk brakes are exposed to the airstream. Therefore, disk brakes are more reliable in hard stops because the brake materials in contact retain most of their sliding friction characteristics. Why so much reliance on the front brakes?

276. Braking

Marieke is driving a car on a level road, puts the car into neutral, and coasts. At the instant of zero velocity, she abruptly applies the brakes. What does she feel? If she drives the car up a gentle slope, puts the car into neutral, coasts, and then applies the brakes at the instant of zero velocity, what does she feel? Are the two cases identical?

Why is the first elevator usually going in the wrong direction? If you are on a low floor, chances are you probably want to go up because there are more possible destinations above you than below you. However, there are also probably more elevators above you than below you, which means that when they reach you, they will be going down.

Absence of evidence is not evidence of absence.

—ANONYMOUS

The Cosmos is about the smallest hole that a man can hide his head in.

—G. K. CHESTERTON

Louis XIV took seriously the obligations implicit in his cosmic title [the Sun King]. He founded a royal observatory in Paris, and here the director, Jean Dominique Cassini, discovered four satellites of Saturn, naming them, predictably, after the King. With Louis's approval Cassini sent Jean Richer to Cayenne, in French Guiana, latitude 5 degrees north, where he found that a pendulum swings more slowly than in Paris, 49 degrees north, which indicated that gravity is less intense near the Equator than in higher latitudes. Since gravity was known to increase as a body came close to the centre of the Earth, it followed that the Earth is not a perfect sphere, but an oblate spheroid.

—Vincent Cronin

277. Car Surprise

Imagine that you have two identical model cars, one black and one white, both having four wheels that roll independently. You lock the front wheels of the white car and the rear wheels of the black car by inserting a piece of folded paper between the wheels and the body. Then you release the cars at the top of an inclined slippery board, front ends pointing downward. What do you predict will happen?

278. Engine Brakes

Some owner's manuals advise drivers to use the car's engine as a "fifth brake" when descending a long, steep grade. In which gear is this braking effect greatest?

279. The Transmission

Steam locomotives and electric cars do not need transmissions, but cars powered by internal-combustion engines do. Why?

280. The Groovy Tire

The purpose of the tread on tires is to increase their grip on the road. If you agree with this statement, then (1) why do drag racers use "slicks" (tires with no tread on them) and (2) why do brake linings have no tread?

281. The Strong Wind

Mr. X is driving fast. A strong wind is blowing from the left, but fortunately the road is dry, so the car has no problem staying in its lane. Suddenly the driver ahead of Mr. X slows down, forcing him to step on the brakes. Mr. X applies the brakes too hard; the wheels lock and slide on the highway. Unexpectedly, the strong wind now easily pushes the car into the next right lane, as if the road had turned into slippery ice. Why is the wind now so effective?

282. Wheels

Compare the wheels shown in the figure. On the bicycle the spokes are mounted tangentially; on the Conestoga wagon they are mounted radially. Why the difference?

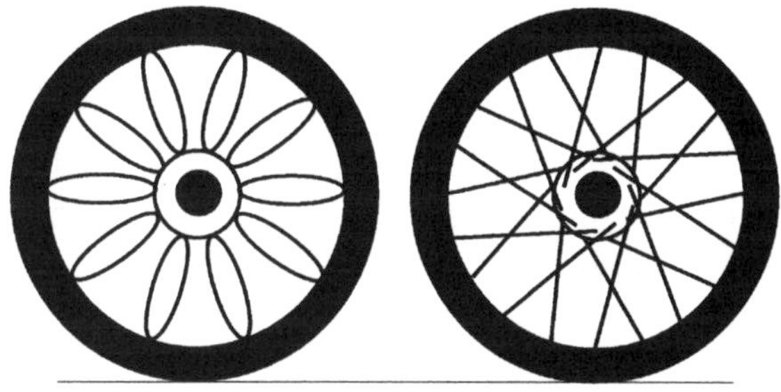

283. Newton's Paradox

Is it true that when a horse pulls forward on a wagon, the wagon pulls backward to the same extent on the horse? In this tug-of-war, it would seem, at least from the point of view of the connecting rope, that it was being pulled by equal forces from each end. In fact, one can show that the forces at the two ends of the rope are always equal and opposite. So, as far as the rope is concerned, the opposing forces always add to zero. Therefore, when starting from rest there can be no ensuing motion. So how does the clever horse succeed in pulling the wagon forward from rest?

284. The Obedient Wagons

Luggage wagons pulling one another in tandem behind a tractor at the airport take a surprising path around a curve. What is the path of each successive wagon? Why?

This most beautiful system of the sun, planets, and comets, could only proceed from the counsel and domination of an intelligent and powerful Being.

—Isaac Newton

Nature and Nature's
Laws Lay hid in Night
God said, Let Newton
be, and all was Light.

—Alexander Pope

Begin with two well-known formulas for uniformly accelerated motion: $v = at$ and $s = 1/2\ at^2$. Solving for acceleration a, and setting the two expression equal to each other, produces $v/t = 2s/t^2$. Simplifying and using $v = s/t$ yields $v = 2v$ or $1 = 2$.

Listen, there's a hell of a good universe next door: let's go!

—E. E. CUMMINGS

285. The Escalator

Consider an escalator, either the kind that ascends or descends between floors or the horizontal type found in airports. As more people get on the escalator, what do you predict should happen to the speed of the escalator? What does happen?

You and I are exceptions to the laws of nature; you have risen by your gravity, and I have sunk by my levity.

—SYDNEY SMITH

286. Roller Coasters

Steven, Annelies, and Annabel enjoy roller-coaster rides. Steven prefers to sit in the front car, Annelies prefers a middle car, and Annabel rides in the last car. As the roller coaster climbs up and over the first hill, why is the ride experienced differently for each of the three passengers?

A man said to the Universe:
"Sir, I exist."
"However," replied the Universe,
"The fact has not instilled in me a sense of obligation."

—STEPHEN CRANE

287. Clothoid Loop

Why are large loops in roller-coaster tracks not circular but teardrop-shaped (clothoid) instead?

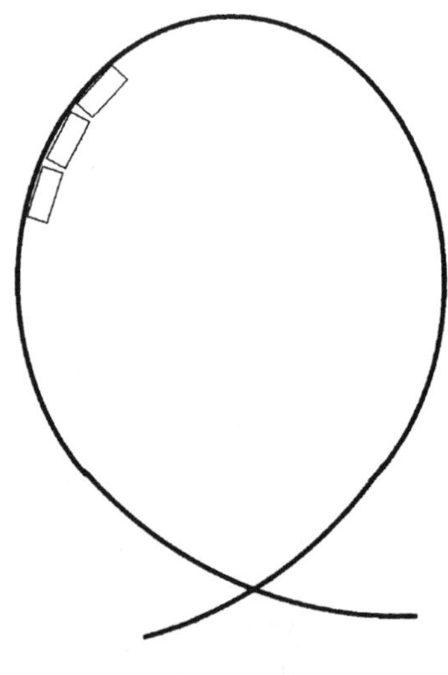

Clothoid loop

288. Turning the Corner

As an automobile turns the corner, the front wheels travel arcs of different radii, and so do the back wheels. Exactly how does the automobile accomplish this feat? Do any of the wheels slip?

289. The Mighty Automobile

An engine of 20 horsepower or less is all that is needed to drive any automobile at a constant 50 miles per hour. Why build cars with 100 horsepower or 200 horsepower, or more?

290. Front-Wheel-Drive Cars

Why are front-wheel-drive cars so successful on snow-covered streets compared to rear-wheel-drive cars? What can one do to improve the traction of a rear-wheel-drive pickup truck?

291. The Hikers

David, Richard, and Paul hike almost every weekend. They pack their backpacks in such a way that the heavier, denser items are nearer to the top and the lighter, less dense items are lower. Does this scheme make sense scientifically? Anatomically?

292. The Fastest Animal Runners

The cheetah is the fastest land animal, able to reach maximum speeds of about 70 miles per hour in short bursts for a few seconds and perhaps maintain a speed of more than 50 miles per hour for ten seconds or so. The pronghorn deer of the Colorado plains is almost as fast, and certainly has better endurance, being able to

I know that this defies the law of gravity, but, you see, I never studied law.

—Bugs Bunny

Science cannot solve the ultimate mystery of Nature. And it is because in the last analysis we ourselves are part of the mystery we are trying to solve.

—Max Planck

The unrest which keeps the never stopping clock of metaphysics going is the thought that the nonexistence of the world is just as possible as its existence.

—William James

If I were granted omnipotence, and millions of years to experiment in, I should not think Man much to boast of as the final result of all my efforts.

—Bertrand Russell

Inquiry into final causes is sterile, and, like a virgin consecrated to God, produces nothing.

—Francis Bacon

While I was gone, somebody replaced all my furniture with its exact replica. When I told my roommate, he said: "Do I know you?"

—Steven Wright

run at about 55 miles per hour for many minutes, quite a long time. The elephant really cannot run at all—that is, get all four feet off the ground simultaneously. How do you explain the differences in their running ability?

293. The Oscillating Board

A board is placed upon the two rotating shafts as shown. Soon one sees this board oscillating left and right repeatedly. Why doesn't it just shoot off in one direction as a projectile?

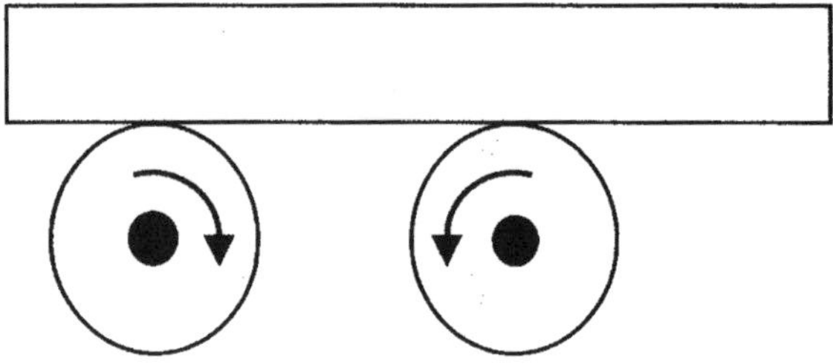

*294. A Cranking Bicycle

A bicycle is being lightly held in a upright position with the cranks (the bars connecting the pedals to the gear) vertical. A horizontal backward pull is applied to the lower pedal. In which direction will (a) the bicycle start to move and (b) the cranks rotate?

*295. Turning a Corner

When a bicycle is suddenly tilting to one side, the cyclist gets out of this predicament by steering into the direction of the fall. In contrast, when a cyclist is about to turn a corner, just before reaching it she will first throw the front wheel over in the opposite direction. Why?

*296. Race Driver

Professional race drivers increase their speed when going around a curve. Why?

*297. The Wall Ahead

You are driving too fast along a road that ends in a T-shaped intersection with a highway. There is a block wall directly ahead on the far side of the highway, and no car is visible in either direction. What should you do to avoid hitting the wall—steer straight at the wall and fully apply the brakes, or turn left into a circular arc as you enter the highway?

Why is it that you physicists always require so much expensive equipment? Now the Department of Mathematics requires nothing but money for paper, pencils and waste paper baskets and the Department of Philosophy is better still. It doesn't even ask for waste paper baskets.

—Anonymous university president

In questions of science the authority of a thousand is not worth the humble reasoning of a single individual.

—Galileo

10 Born to Run

MOST SPORTS COMBINE MECHANICS WITH the biophysics of the human animal. Therefore, each of us is limited by the laws of physics and by our physiological design. A muscle's strength is proportional to its cross-sectional area, but big muscles without the proper skills do not lead to world-class performance. With training and practice we can increase our abilities toward maximum performance. So let your mind tackle these challenges, which illustrate how remarkable some human performances can be.

298. Strong Women

Kilogram for kilogram of lean body weight, women are as strong as men. True or false?

299. Hanging in the Air!

Great jumpers on the basketball court sometimes seem to hang in the air with exceptional body control before taking a remarkable shot. But this effect seems even more amazing on the ballet stage, where one witnesses remarkable, seemingly effortless jumping ability combined with exceptional body control and grace. Ballet dancers appear to be able to willfully suspend their bodies in flight for several seconds. Can an athlete really "hang in the air"?

300. Good Running Shoes

A great variety of running shoes have become available in the past fifteen years, some with air pockets and foam wedges and some without a tongue. Are most shoe design features simply commercial hype, or is there some real biophysics behind the running shoes now available?

301. Sprinting

In short races, 100 meters or less, why is breathing during the race not necessary?

302. Long-Distance Running Strategy

Why do runners in middle- and long-distance races—1,500 meters and up—avoid running at their maximum speeds in the early stages of the race? Surely, one

Scientists leave their discoveries like foundlings on the doorstep of society, while the stepparents do not know how to bring them up.

—Alexander Calder

Here, Kitty, Kitty!
H. H. Hetherington added as co-author one F. D. C. Willard (Felix Domesticus Chester Willard)[1], a collaborator whose contribution to the research was probably rather indirect.

[1]J. H. Hetherington and F. D. C. Willard, "Two-, Three-, and Four-Atom Exchange Effects in bcc ^{3}He," *Phys. Rev. Lett.* 35, 1442–1444 (1975).

Robert Adair's book *The Physics of Baseball* sets the conditions that would alter the flight of a 400-foot home run to dead center:

- A 1,000-foot increase in altitude adds 7 feet to the shot.
- A 10-degree increase in air or ball temperature boosts it by 4 feet.
- A 5-mile-per-hour tailwind adds 15 feet.
- A pitch that's 5 miles per hour faster ends up 3.5 feet farther from home plate.

should run at the maximum speed all the way through the race to maximize one's performance instead of putting on a burst of speed near to the end.

SMITH'S LAW OF INERTIA
A body at rest tends to watch television.

Tendons, though their strength is less than that of mild steel (≈450 megapascals), are about seven times less dense and so are actually stronger pound for pound.

To move forward, you must subject yourself to instability.

—ANONYMOUS

Even though your brain makes up only about 2 percent of your body's total weight, it uses about 25 percent of your total blood supply. In children, the brain sometimes uses up to 50 percent of the total blood supply.

Proof by induction is not as prevalent as proof by intimidation.

—AUSTIN TRAIN

303. Location Effects on High-Jump Records

Since Newton's time, we have known that the effective value of the acceleration of gravity *g* depends upon both the altitude and the rotation of the Earth at a particular latitude. In fact, there is a well-known expression for calculating *g* for any given latitude and altitude. Then why doesn't the committee that verifies world records in track and field take the geographical location into account, particularly for the high jump and the long jump?

304. High-Jump Contortionist

High jumpers use the "Fosbury flop," twisting so that their backs are downward when they go over a bar placed much higher than their own heights. Why do they arch their bodies so much at the apex of the jump? You would think that the extra effort to then flip their legs over the bar could have been used to jump higher! Can the center of gravity of the high jumper pass below the bar?

305. Pole Vaulter

In pole vaulting, the object is to clear the highest bar placement. The present pole vault record is more than 20 feet. Shouldn't the vaulter simply choose the longest pole of the best material (i.e., greatest elasticity) and do the pole vault, all other factors remaining the same as for previous vaults?

306. Basketball

Why is backspin so important in shooting a basketball? Every player practices shooting the basketball from the fingertips with a slight flick of the wrist to automatically put backspin on the ball.

307. Doing the Impossible!

Determine whether you can do this stunt. Face the edge of an open door with your nose and stomach touching the edge and your feet extending forward slightly beyond it. Now try to rise on tiptoes. Why is this feat impossible?

308. Reaction Time with a Bat

In baseball, the pitcher begins 60 feet, 6 inches from home plate, but the ball is released about 3 feet closer. At 90 miles per hour (132 feet per second) the baseball arrives above home plate pretty quickly. When should most batters begin their swing of the bat?

309. Can Baseballs Suddenly Change Direction?

Most professional baseball players insist that they have seen a pitched ball travel in a straight line, then curve suddenly just before reaching home plate. Can this behavior be true? How?

310. The Curveball

What makes a curveball pitch curve? In what direction does the ball curve for a right-handed pitcher? For a left-handed pitcher?

All the arts and sciences have their roots in the struggle against death.

—St. Gregory of Nyssa

The human hand possesses twelve hinge joints and five universal joints, allowing a total of 22 degrees of freedom of movement.

QED = quite easily done

Science is the belief in the ignorance of experts.

—Richard P. Feynman

Astrology is a science in itself and contains an illuminating body of knowledge. It taught me many things, and I am greatly indebted to it. Geophysical evidence reveals the power of the stars and the planets in relation to the terrestrial. In turn, astrology reinforces this power to some extent. This is why astrology is like a life-giving elixir for mankind.

—Albert Einstein

Above the speed of about 2.3 meters per second, running uses less energy than walking.

An experimentalist observes what can't be explained, and a theoretician explains what can't be observed.

—Anonymous

It's okay to sleep with a hypothesis, but you should never become married to one.

—Anonymous

The elastic recoil of running shoes returns from 54 to 66% of the strain energy—a poor second to the arch of the foot, which returns 80 percent of the energy stored in it.

The human body loses about 2 pounds of cells each day (mostly from the skin and the lining of intestines).

[Science proceeds by] putting nature on the rack and torturing the answers out of her.

—Attributed to Francis Bacon

311. Scuffing the Baseball

What exactly does the pitcher gain by scuffing a baseball—that is, by roughing up a spot on the surface of the ball?

312. Watching the Pitch

Some of the best baseball batters claim that they can watch the spin on the ball from its release point from the pitcher's hand to when the ball strikes the bat. What do you think?

313. The Bat Hits the Baseball

The speed at which the ball leaves the bat in baseball depends upon the collision location along the bat. For the maximum batted-ball speed, is the best location at the center of percussion of the bat, that point on the bat that transfers no momentum from the ball collision to the handle and vice versa?

314. Underwater Breathing

Breathing underwater at a depth of 2 meters (about 6 feet, 8 inches) is usually done with scuba equipment. Why not simply breathe through a long tube or snorkel whose upper end is attached to a float at the surface?

315. Springboard Diving Tricks

After an expert diver jumps off a springboard, can she start to somersault and twist in midair long after she has left the board? Or must the somersaulting and twisting begin before leaving contact with the board?

316. Cat Tricks

If you drop a cat upside down above a soft cushion, the cat will mysteriously land on its feet. How can the animal achieve a net rotation in space without anything to push against?

317. Astronaut Astrobatics

Can an astronaut who is motionless—that is, has no angular momentum initially—reorient herself in any direction she wants?

318. The Feel of the Golf Shot

Some professional golfers are aware of the moment the golf club head strikes the ball by the feel they experience at their hands. Does this sensation occur during the contact with the ball?

319. Skiing Speed Record

The speed record for downhill skiing is about 3 miles per hour *faster than* the maximum downward falling speed for the same human through the air—that is, the maximum attained terminal speed through the air while falling. What is the physics here?

320. "Skiers, Lean Forward!"

Why do ski instructors call out to learners, "Lean forward"? This expected body orientation is unnatural to beginners, most of whom try to remain parallel to the trees. Is the instructor's advice good physics?

Nature uses as little as possible of anything.
—Johannes Kepler

When asked what characteristic physics Nobelists had in common, Enrico Fermi said after a long pause, "I cannot think of a single one, not even intelligence."

The maximum force that a muscle can exert is about 50 lb/in^2 or 35 N/cm^2 in human beings.

Talent does what it can,
Genius does what it
must.

—Anonymous

Get your facts first, then you can distort them as much as you please.

—Mark Twain

Given the identical conditions of a warm and humid environment, on the average during a 40-minute period a woman will sweat 400 milligrams compared to 600 milligrams for a man.

Bodily exercise, when compulsory, does no harm to the body; but knowledge which is acquired under compulsion obtains no hold on the mind.

—PLATO

The center of gravity of a person standing erect lies on a vertical line that touches the floor 3 centimeters in front of the ankle joint.

The power of the heart in a normal adult at rest is about 1 watt, which is less than 1 percent of the metabolic rate.

The skin temperature of a nude person sitting in a room at 22 °C is on the average about 28 °C.

A sneeze comes out your nose at 180 miles per hour.

The more questions we answer, the more answers we end up questioning.

—ANONYMOUS

321. Ski Slope Anticipation

Why do professional skiers "prejump"? Just before reaching the edge of a steeper section of the slope, the skier quickly rises from the crouch position and pulls her legs upward to make the skis leave the ground before reaching the steeper part. Is there any advantage to this technique?

322. Riding a Bicycle

Why is it easier to ride a bicycle than to run the same distance?

323. Give Me a Big V

Is there any physical advantage to the V formation often assumed by a flock of migrating birds?

324. Deadly Surface Tension

Surface tension is a force that is hardly noticeable to large animals and yet is deadly to insects. Why?

*325. Animal Running Speeds

The maximum running speed on level ground is almost independent of the size of an animal. For example, a rabbit can run as fast as a horse. However, in running uphill the small animals easily outpace larger ones. A dog runs uphill more easily than a horse, which must slow its pace. What is the dimensional argument that agrees with these findings?

*326. Scaling Laws for All Organisms

The amount of energy required to sustain life in all organisms—that is, the metabolic rate—is roughly proportional to the body mass raised to the power 3/4. Shouldn't an organism's energy requirements grow in direct proportion to body mass itself, not as some fractional power less than 1?

*327. Tennis Racket "Sweet Spot"

Why does a tennis racket have a "sweet spot"? Where is it located? Can there be more than one "sweet spot"?

*328. Golf Ball Dimples

Why are there dimples on golf balls? Surely they must increase the turbulence around the ball in flight!

A sign on the wall of a graduate student research laboratory: THEY CAN'T FIRE ME: SLAVES HAVE TO BE SOLD

In the absence of any noticeable perspiration there is an insensible evaporation of water from the skin and lungs of the human body that amounts to 600 grams of water per day.

I believe a leaf of grass is no less than the journeywork of the stars . . . and the running blackberry would adorn the parlors of heaven.

—Walt Whitman

A 70-kilogram (154-pound) man normally uses about 10^7 joules a day. His average metabolic rate is about 120 watts. It falls to 75 watts while sleeping and rises to 230 watts while walking.

Cheetahs can accelerate from 0 to 40 miles per hour in 2 seconds.

11 Third Stone from the Sun

EARTH IS OUR MIRACULOUS PLANET. WE BASK in the sunlight that passes through its atmosphere, we swim in the waters of its lakes and oceans, we trudge through the snow and cold winds in the winter, and we send signals to each other via radio waves in the atmosphere. But do we know how to use physics to explain these phenomena? Here is a small sampling of questions to whet your appetite for a more thorough appreciation of how this big dynamical system on our planetary island in the cosmos operates. If you have learned how to apply many of the concepts from previous chapters, you should be ready for these challenges.

329. California Cool

In the United States, Pacific Coast water is usually much colder than Atlantic Coast water. Why?

330. Waves at the Beach

An observer on the beach always sees larger waves come in directly toward him, with the wave crests parallel to the shore, even though some distance out from shore they are seen to be approaching at an angle. What makes the waves straighten out?

331. Ocean Colors

From a plane flying above the ocean, the water looks much darker directly below than toward the horizon. Why?

332. Stability of a Ship

Generally, we associate a low center of gravity with stability. However, for a floating ship, the center of gravity must be above its center of buoyancy (where the upward buoyant force may be considered to be centered) to ensure stability. Why?

333. Longer Ships Travel Faster

A 100-meter-long ship reaches full ("hull") speed at about 28 miles per hour, whereas a 10-meter-long ship finds it difficult to exceed 8 miles per hour. A duck (think of it as a very short ship!) can actually swim several times faster fully submerged than on the surface. Why is it that longer ships can travel faster?

The Time Service Department of the U.S. Naval Observatory in Washington, D.C., keeps the most precise atomic clock in the U.S. By dialing 900-410-TIME, one can hear an announcement accurate to one-tenth of a second. Clock watchers with a shortwave radio can tune in to station WWV in Fort Collins, Colorado (2.5, 5, 10, 15, and 20 megahertz), for the Bureau of Standards time signal.

The more important fundamental laws and facts of physical science have all been discovered and these are now so firmly established that the possibility of their ever being supplanted in consequence of new discoveries is remote.

—Albert Michelson (ca. 1890)

Due to the slowing rate of the Earth's rotation the twentieth century was about 25 seconds longer than the nineteenth century.

One of the most active regions on earth in terms of thunderstorm activity is Java, where thunder is heard 223 days per year. In the U.S., central Florida has the highest number of thunderstorm-days per year with 90. The lowest number occurs along the Pacific coast region of northern California, Oregon, and Washington, where thunderstorms and lightning are rare.

—Martin A. Uman

The Republic has no need for men of science.

—Jean-Paul Marat (on the occasion of condemning the chemist Lavoisier to death at the guillotine)

Do not seek to follow the footsteps of the men of old; seek what they sought.

—Matsuo Basho

Thunder cannot usually be heard if the lightning causing it is more than about 15 miles away.

334. Polar Ice

Why does Antarctica have eight times as much ice as the Arctic?

335. The Arctic Sun

The diagram shows the successive positions of the sun during a period of a few hours, as observed in Alaska. Can you tell approximately which compass direction the observer was facing? Roughly at what time of day or night was the lowest elevation of the sun observed?

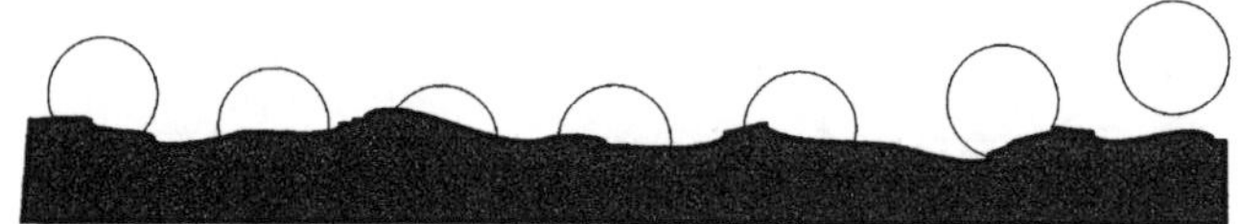

336. Circling Near the Poles

Lost polar explorers reportedly have a strong tendency to circle steadily toward the right near the North Pole and to the left near the South Pole. Can you think of a possible explanation?

337. Weather Potpourri

Do you agree with the following homespun weather predictions? If you do, what is the scientific basis for them?

1. Your joints are more likely to ache before a rainstorm.
2. Frogs croak more before a storm.
3. If leaves show their undersides, rain is due.
4. A ring around the moon means rain if the weather has been clear.

5. Birds and bats fly lower before a storm.
6. You can tell temperature by listening to a cricket.
7. Ropes tighten up before a storm.
8. Fish come to the surface before a storm.
9. "Singing" telephone wires signal a change in the weather.

338. Wind Directions

Winds on the Earth blow directly from higher-pressure areas to lower-pressure ones. True or false?

339. Deep Freeze

For purely astronomical reasons, the Southern Hemisphere of the Earth should suffer colder winters and hotter summers than its northern counterpart. In fact, the lowest temperature ever recorded, –128.6 °F (–89.2 °C), occurred in the Antarctic. However, by and large the peculiar conditions existing in the Southern Hemisphere compensate for this trend very effectively. What mysterious astronomical reasons and peculiar conditions are we referring to?

340. Weather Fronts

When cold and warm air lie alongside each other, as they do in a weather front, even if no pressure difference exists at ground level, the warm air and the cold air will act as high- and low-pressure zones, respectively. The pressure difference between them then gives rise to the so-called thermal winds. On the other hand, we know that cold air is denser than warm air, so it seems that it is the cold air that should be associated with a high-pressure zone. How do we resolve this apparent contradiction?

Los Angeles is moving north toward San Francisco at the rate your fingernail grows.

Before a 1,000-ton ship can float, somebody has to misplace 1,000 tons of water. This is the captain's job.

—From Art Linkletter's *A Child's Garden of Misinformation*

The highest officially recorded sea wave was calculated at 112 feet from trough to crest; it was measured during a 68-knot hurricane by Lt. Frederic Margraff (U.S. Navy) from the U.S.S. *Ramapo,* traveling from Manila, Philippines to San Diego, CA on the night of February 6–7, 1933.

The Guinness Book of World Records (1998)

Weathersfield, Connecticut, is the only town in the United States hit twice in a row by a meteor. A possible explanation: Weathersfield is close to Hartford—the insurance center of America.

Because of the air currents, it often snows upward at the top of the Empire State Building in New York.

The global average surface temperature is 14.0 °C. The Northern Hemisphere is a bit warmer than the Southern, averaging 14.6 °C versus 13.4 °C. Over the period 1861–1997 the average global temperature rose 0.57 °C. Much of the net warming is at night; for the period 1950–1993, nighttime average minimum temperatures increased 0.18 °C per decade, while daytime average high temperatures increased only 0.08 °C per decade.

Spring, the time to plant crops, moves 1 degree northward in 4 days and arrives 1 day later for every 100 feet of altitude.

The record rainfall for a short duration is 12 inches in 42 minutes at Holt, Missouri, in 1947.

341. Lightning and Thunder

Thunder is the sound created by rapidly expanding gases along the channel of a lightning discharge. But if the lightning stroke is practically instantaneous, why does thunder sound the way it does? Why does it rumble, roll, peal, and clap?

342. Lightning without Thunder?

Can there be lightning without thunder?

343. Direction of the Lightning Stroke

Does lightning between cloud and ground travel upward or downward?

344. Outdoor Electric Field

When you step outdoors on a clear day you are surrounded by a downward electric field of about 100 volts per meter at the Earth's surface. The field intensity varies considerably with location, topography, the hour of the day, and the state of the weather. Readings made at the top of mountains average much higher than those made at sea or on flat land. Conversely, observations made in valleys average somewhat lower. When a charged thundercloud comes along, the field may go up to 10,000 volts per meter. Why doesn't this voltage kill you?

345. Negative Charge of the Earth

Why is the Earth's surface negatively charged?

346. Peak in the Global Electric Field

The variation of the global atmospheric electric field shows a daily maximum at 1900 (7:00 P.M.) Universal time (Greenwich mean time). Any idea why?

347. Radio Reception Range

It is much easier to pick up distant AM (and shortwave) radio signals at night. In fact, to prevent interference, most AM stations are required to cut their power or even to leave the air at dusk. What conditions exist at night that help to increase the range of radio waves?

348. Car Radio Reception

You may have noticed that while the AM band on the car radio cuts out when you are passing under a bridge, the FM band in the same situation would continue to play. Why is there such a big difference in the reception of AM and FM signals?

349. Magnetic Bathtubs

Every stationary iron object in the United States is magnetized with a north pole at the bottom and a south pole at the top! This includes bathtubs, filing cabinets, refrigerators, and even umbrellas with an iron shaft left standing for a while. Any ideas why?

350. The Bathtub Vortex

When the water in a bathtub is allowed to drain, it develops a vorticity or swirling motion around the

On the average, a given commercial airplane is struck by lightning once in every 5,000 to 10,000 hours of flying time. Aircraft struck by lightning almost always continue to fly. Generally, lightning leaves pit marks or burn marks on the aircraft's metallic skin or burn or puncture holes through it. The FAA reports hole diameters up to 4 in., a common size being ½ in.

—Martin A. Uman

Coriolis Effect? Guests at cocktail parties tend to circle clockwise around the buffet table. It's a high-pressure area!

The difference between high- and low-pressure areas is generally less than 3 percent.

I am tired of all this thing called science.... We have spent millions on that sort of thing for the last few years, and it is time it should be stopped.

—Simon Cameron (U.S. senator from Pennsylvania, 1861)

How realistic are physics problems involving deep holes in the earth? For example, one can show that an object moving in a smooth, straight tunnel dug between two points on the earth's surface under the influence of gravity alone will execute simple harmonic motion with a period of 84.3 minutes. The problem is that you just couldn't hope to keep a hole open much deeper than, say, 16 kilometers. At about 30 kilometers, for instance, the pressure and temperature are so great that even the pores and cavities of solid rock close.

Meteorologists don't age, they just weather away.

All of the earth's water would fit into a 700-mile cube.

About seven out of ten red sunsets are predictive of good weather in the northern climates.

—Gary Lockhart

drain. Many people believe that the rotation of the vortex is always counterclockwise in the Northern Hemisphere and always clockwise in the Southern Hemisphere, and that the effect is due to the rotation of the Earth. Is this belief justified?

351. Gravity Near a Mountain

You might expect the gravitational attraction due to a nearby mountain range to cause a plumb bob to hang at an angle slightly different from vertical. This example actually appears in many physics textbooks. However, the observed deflection is, surprisingly, much smaller than what is predicted by theoretical calculations. In fact, the deflection is practically zero, apparently implying that a mountain range exerts no extra pull on a plumb bob. Can you see a way out of this seeming paradox?

352. Gravity inside the Earth

Many people are under the impression that the gravitational field strength $g(r)$ decreases as one goes down from the Earth's surface. For a solid sphere of radius R, total mass M, and uniform density, the gravitational field at a distance r from the center is $g(r) = (GM/R^3)r$,

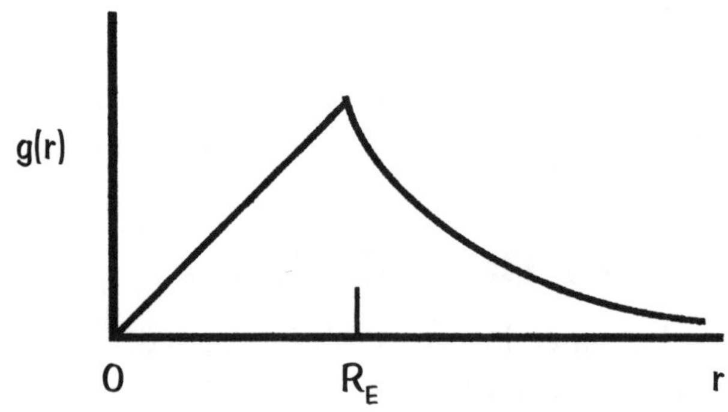

showing a linear increase from the center to the surface. Can this simple relationship be expected to hold in the real world?

353. Why Is Gravitational Acceleration Larger at the Poles?

It is often stated that the gravitational acceleration at the poles is larger than at the Equator because the surface of Earth at the poles is some 21 kilometer closer to the center of the Earth due to its flattening. Is this the main reason?

354. The Green Flash

One can sometimes glimpse an extraordinary effect called the green flash at sunset. Just as the last of the solar disk is about to disappear, for several seconds it turns a brilliant green. The effect can be seen only if the air is clear and the horizon is distinctly visible—usually at sea or in mountain or desert country. How does nature produce this phenomenon?

*355. Meandering Rivers

There is no such thing as a straight river. In fact, it was found that the distance any river is straight does not typically exceed ten times its width at that point. At first we might suppose that a river twists and bends in direct response to peaks and dips in the landscape. Not at all! On a typical smooth and gentle slope, water does not flow straight downhill; it winds and turns as if desperately trying to avoid the straight path to the bottom. Why?

Men are struck by lightning four times more often than women.

Mount Everest is said to be the tallest mountain, at 29,028 feet above sea level. But another way of measuring mountain peaks is by their distance from the center of the earth. On this basis, Equador's Chimborazo, at 20,561 feet above sea level, would be taller than Everest by a whopping two miles. This is because the earth bulges at the equator (near Chimborazo) and flattens toward the poles.

Most of the Atlantic is somewhat below sea level.

—From Art Linkletter's *A Child's Garden of Misinformation*

Suicides rise by 30 percent whenever pressure changes by more than 0.35 inch in a day.

*356. Energy from Our Surroundings

There is a widespread belief that because of the second law of thermodynamics we cannot use the energy in our surroundings to do useful work. For example, a powerboat cannot pump in water, extract energy from it to drive its propellers, and throw overboard the resulting lumps of ice. The second law seems to forbid such possibilities because of the lack of a suitable heat reservoir at a low temperature. In fact, such a heat reservoir does exist and is readily available. Any ideas?

There have been substantial changes in the solar power output over the centuries, and they seem to correlate with temperature changes on earth. For example, the period of lowest solar activity occurred between 1600 and 1700, when sunspots practically vanished during the latter half of the century. That was also the coldest period in the past thousand years, sometimes called "the little ice age."

*357. Temperature of the Earth

What determines the temperature of the Earth? It cannot be the heat trickling up from the interior of the Earth. Its amount is too negligible compared to the solar radiation absorbed by the Earth's surface. At equilibrium, the amount of sunlight absorbed must equal on average the amount of energy radiated back into space. The equilibrium temperature obtained using this equality is 256 K, or –17 °C, some 30 °C below its actual measured value. Did we make a mistake, or is there something important we left out?

Sheridan's Rhyming Calendar

Autumn months: wheezy, sneezy, freezy
Winter months: slippy, drippy, nippy
Spring months: showery, flowery, bowery
Summer months: hoppy, croppy, poppy

*358. The Greenhouse Effect

Is it reasonable to describe the connection between increasing concentrations of carbon dioxide and the presumed rising global temperatures as the "greenhouse effect"? Some people claim greenhouses are warm because of radiation trapping: the glass is transparent to solar radiation but opaque to infrared. Others say greenhouses are merely shelters from the wind—all they do is suppress convective heat transfer. Who is right?

*359. Measuring the Earth

In about 200 B.C. Eratosthenes, director of the great library in Alexandria, came upon a simple method of determining the circumference of the Earth. He read that in Syene (Aswan), Egypt, at noon on June 21, obelisks cast no shadows and sunlight fell directly down a well. He observed that in Alexandria (located directly north of Syene) at noon on the same date the sun was about 7 degrees south of the zenith. He next had the distance between Syene and Alexandria determined, probably by a *bemetatistes,* a surveyor trained to walk in equal paces. The distance came to 5,000 stadia. Using this figure he calculated the Earth's circumference to be (360°/7°) × 5000, or roughly 250,000 stadia, equivalent to 42,000 to 46,000 kilometers, about 5 percent too large.

Although the method is simple, it is laborious. Today anyone can determine the size of the Earth to within 10 percent by simply watching a sunset. Can you explain how?

A 1964 study of Wisconsin schoolchildren showed that students were more quiet on clear days and more restless on cloudy days. Test scores were highest during the times of greatest restlessness.

—GARY LOCKHART

On a normal day, a cubic centimeter of air contains 1,200 positive ions and 1,000 negative ions. These negative ions are generally oxygen with an extra electron, and the positive ones are carbon dioxide minus an electron.

The real voyage of discovery consists not in seeking new lands but seeing with new eyes.

—MARCEL PROUST

12 Across the Universe

THERE IS A WHOLE UNIVERSE OF PHYSICS CHALLENGES out there, but in this chapter we do not need to venture very far beyond own Solar System with our single Sun to find surprises. We all have enjoyed the stars of our galaxy as a backdrop for the planets that wander across the sky. Our Moon, of course, is a regular visitor to our heavens, both night and day. In the twentieth century artificial Earth satellites joined the crowd, scurrying quickly across the sky to remind us how close they really are. With so many familiar objects, we could ask an infinity of questions. Here are a few to challenge your thinking.

360. Visibility of Satellites

Why is it that most artificial Earth satellites can be seen only during roughly the two hours after sunset or the two hours before sunrise?

361. A Dying Satellite

A dying artificial Earth satellite makes its terminal appearances at the same time and in the same part of the sky for several days before disintegrating in the atmosphere. Why?

362. Cape Canaveral

Why were the first American satellites launched from Cape Canaveral, Florida? More generally, why do space launch sites, such as the Kennedy Space Center at Cape Canaveral, tend to be located toward the tropics?

363. Weightlessness in an Airplane

Weightlessness can be achieved for 20 to 30 seconds in an airplane executing one of the following maneuvers: (a) inside loop (with the center of the loop above the plane); (b) outside circular loop (with the center below the plane); (c) outside parabolic loop. Which one?

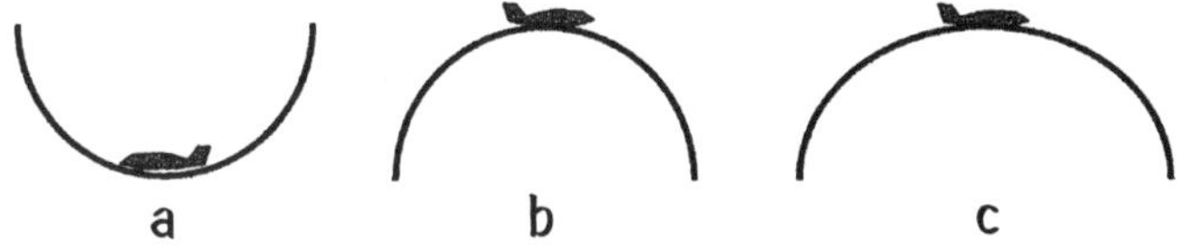

364. A Candle in Weightlessness

Will a candle burn in weightlessness?

I am much occupied with the investigation of the physical causes [of the motions of the Solar System]. My aim is to show that the heavenly machine is not a kind of divine, live being, but a kind of clockwork . . . insofar as nearly all the manifold motions are caused by a most simple, magnetic, and material force, just as all motions of a clock are caused by simple weight.

—JOHANNES KEPLER

The Universe is made of stories, not of atoms.

—MURIEL RUKEYSER, POET

Before Tycho Brahe, the best measurements in astronomy had inaccuracies of at least 10 arc-minutes. Brahe's measurements had inaccuracies of only 2 arc-minutes.

The sun is 600,000 times as brilliant as the full moon.

Describe the Universe and give two examples.

—FINAL EXAM IN ASTRONOMY

Should we not attribute to God, the Creator of nature, that skill which we observe in common makers of clocks? For they carefully avoid inserting in the mechanism any superfluous wheel, or any whose function could be served better by another slight change of position.

—GEORGE RHETICUS (AN EARLY SUPPORTER OF COPERNICUS'S IDEAS)

[In this paper it has been shown] that a rocket capable of carrying a man to the Moon and back would need to be of fantastic size and weight—so large, indeed, that the project could be classified as impossible.

—J. HIMPAN AND R. REICHEL, *AMERICAN JOURNAL OF PHYSICS* 17, 251 (1949)

For every 7.9 kilometers you go on Earth, the surface dips 4.9 meters, the height that a body falls in 1 second. Hence the orbital speed in the lowest Earth orbit is 7.9 kilometers per second.

365. Boiling Water in Space

An astronaut in a spacecraft puts a kettle of water on an electric stove to boil under conditions of weightlessness. When he checks the kettle an hour later, the water on top is still cold. Why?

366. Maximum Range

You wish to launch a spacecraft so it will reach as far out as possible into the Solar System. Which gives you the maximum range for less fuel—to launch the spacecraft in the direction of Earth's orbital speed when the Earth is closest to the Sun or farthest from the Sun?

367. Air Drag on Satellites

What is the effect of air drag on a satellite traveling through the upper layers of the atmosphere—to slow the satellite down or to speed it up?

368. Separation Anxiety

When a satellite separates from the launching rocket that was used to put it into orbit around the Earth, the rocket is usually seen to overtake the satellite gradually, even though its engine has been shut off. Any ideas why?

369. Changing the Orbit—Radial Kick

The engines of a spacecraft in a circular orbit about the Earth are fired briefly to give the craft an outward radial thrust as shown on page 147 in (a). Will this thrust produce orbit (b) or orbit (c)?

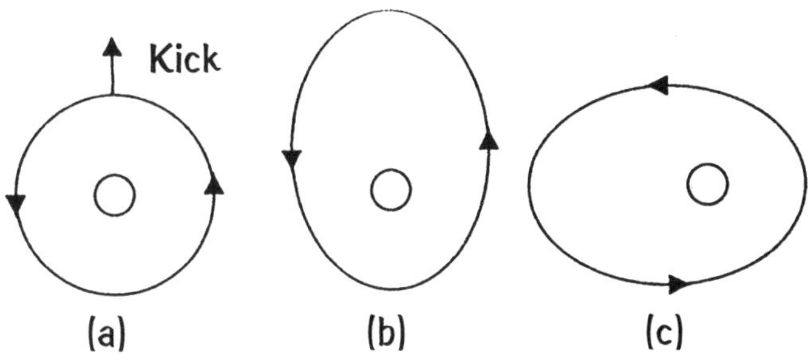

370. Changing the Orbit—Tangential Kick

A spaceship in a circular orbit around the Earth briefly applies a small tangential thrust as shown in (a). Will this thrust result in orbit (b) or orbit (c)?

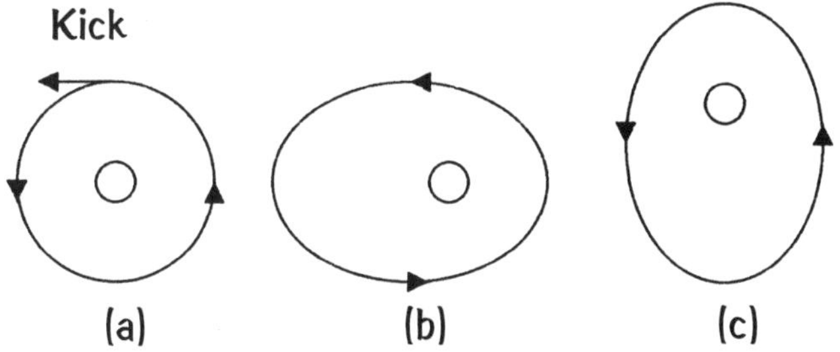

371. Exhaust Velocities

Imagine a rocket moving parallel to the ground high above the Earth. Is it possible for the exhaust gases to move in the same direction as the rocket with respect to the ground and still accelerate the rocket forward?

372. Liftoff Position

During liftoff, shuttle astronauts typically assume a prone (i.e., parallel to the ground) position. Why is this orientation preferable to a sitting-up position?

After Neil Armstrong and Buzz Aldrin in the lunar module *Eagle* had landed on the Moon's Sea of Tranquillity on July 20, 1969, Aldrin opened a miniature Communion kit prepared by his Presbyterian pastor, ate a tiny Host and drank the wine, and silently gave thanks for the intelligence and spirit that had brought the astronauts to the Moon. So much for the separation of church and space!

The space shuttle often flies upside down and backwards, with the reinforced underbelly acting as a shield. The reason for the upside-down position is so the overhead windows can look out on the Earth.

A typical spacesuit weighs about 275 pounds and, like an inflated tire, is pressurized to 4.3 pounds per square inch.

Everyone is a moon, and has a dark side which he never shows to anybody.

—Mark Twain

In a typical space shuttle flight the sky darkens very quickly, becoming totally black by about 2 minutes into the flight.

Robert Edwards found that low points in stock prices occurred at the new and full Moons. High prices tended to occur at the first and third quarters. The difference between these high and low points is only about 1 percent.

—Gary Lockhart

Weightless Sleep

Astronauts' arms float in front of their bodies when they are asleep.

As I lie there in this capsule, looking at the switches, buttons, and readouts, during the countdown I think, "Just think, this thing was built by the lowest bidder."

—Walter Schirra

Pluto, with a diameter of 2,300 kilometers, is smaller than our Moon, and also smaller than the four Galilean satellites of Jupiter, than Titan, a satellite of Saturn, and Triton, a satellite of Neptune.

373. Escaping from Earth?

If a rocket is launched vertically upward with a speed of 11.2 kilometers per second and then the engine is shut off, it will be able to escape from Earth. Now suppose that the rocket is launched almost horizontally with the same initial speed. Neglecting air effects, will the rocket still be able to escape from Earth?

374. Orbit Rendezvous

Imagine that you are the commander of a shuttle on a mission to rendezvous with a space station. The station is at your exact altitude and 50 kilometers ahead in a circular orbit. To close the distance, you fire your thruster rockets to increase the shuttle speed in the direction of the space station. Will this maneuver work?

375. Shooting for the Moon

The Kennedy Space Center at Cape Canaveral has a very favorable location as a launching site due to its proximity to the Equator. Even more interestingly, there is something special about its latitude of 28.5 degrees, which gave the U.S. Apollo program (1966–1972) a competitive edge. This latitude of 28.5 degrees is perfect for launching lunar missions. Can you see why?

376. Rocket Fuel Economy

Which is more economical for a two-stage rocket—that is, which sequence of operations will bring a payload to the greatest height? (a) fire the upper stage after its booster has carried it to its maximum height, or (b) fire the upper stage at a low altitude immediately following the burnout of its booster's propellant? Assume each stage has the same burnout speed and that the acceleration of gravity is the same at all heights.

377. Speed of Earth

When is the Earth moving fastest around the Sun? When is it moving slowest?

378. Earth in Peril?

Is the Earth in any danger of falling into the Sun?

379. The Late Planet Earth

If the Earth were suddenly stopped in its orbital movement, how long would it take to fall into the Sun?

380. Brightness of Earth

Venus and Earth are about the same size. However, viewed from Venus, Earth at its best would appear about six times brighter than Venus ever appears to the Earth. This result occurs despite the fact that Earth is farther away from the Sun and the visible light reflectivity of Venus is greater than that of Earth! How can you explain the apparent paradox?

THE ROCKET CHALLENGE

Except for the hypothetical nuclear fuel, there is no propellant yet known that has enough chemical energy to lift its own weight into orbit, let alone escape Earth's gravity completely. Manned flights are limited to altitudes ranging roughly from 100 nautical miles to 300 nautical miles. Altitudes below 100 nautical miles are not possible because of atmospheric drag, and the Van Allen radiation belts limit manned flights to altitudes below about 300 nautical miles.

Typically, space shuttles fly at an altitude of about 185 miles.

The maximum speed of meteors entering the atmosphere is about 72 kilometers per second.

UFOs are better explained in terms of the unknown irrationalities of terrestrial beings rather than by any unknown rationalities of extraterrestrial beings.

—RICHARD FEYNMAN

Early observers of the moon, including Galileo, noticed that the full moon appeared to be flat. If a round object such as a balloon or ball is illuminated like the full Moon, a distinct three-dimensional effect occurs: a gradual darkening around the edges with the brightest area in the center. That the full Moon has almost equal illumination everywhere across its surface is the result of its unusual surface texture. Light from the Sun striking the Moon is almost completely absorbed by the surface. Only about 12 percent of the sunlight is reflected. In contrast, Earth reflects about 37 percent of sunlight.

—Kim Long

Black holes are where God divided by Zero.

—Steven Wright

We are all in the gutter but some of us are looking at the stars.

—Oscar Wilde

381. Meteor Frequency

On any clear night a meteor can be seen in the sky about every ten minutes. However, their numbers increase toward morning. Why?

382. Slowly Rotating Earth

The planets of our Solar System show a very interesting relationship between mass and period of rotation. In general, the greater the mass, the faster the speed of rotation. Thus Jupiter, with the largest mass of any of the planets, also has the fastest speed of rotation and the shortest period, 9 hours, 50 minutes. Saturn, with a smaller mass, rotates in 10 hours, 14 minutes. Uranus and Neptune, with masses still smaller, rotate in 16 or 17 hours. Finally, Mars, which is far smaller than any of the giant planets, rotates in 24 hours, 37 minutes. However, the Earth is ten times as massive as Mars, yet it rotates in about the same time. Why does the Earth rotate so slowly?

383. Can the Sun Steal the Moon?

If a body is more than 259,000 kilometers from earth, it is attracted more strongly by the Sun than by the Earth, as one can verify by using the inverse-square law of universal gravitation. The average distance from Earth to the Moon is 384,400 kilometers, a much larger distance than 259,000 kilometers; therefore the Moon is pulled more by the Sun than by the Earth—in fact, more than twice as much. Why doesn't the Sun steal the Moon from the Earth?

384. Moon's Trajectory around the Sun

The diagram shows a section of Earth's orbit around the Sun, with the Moon's trajectory around the Earth. Apart from not being drawn to scale, is there anything basically wrong with the diagram?

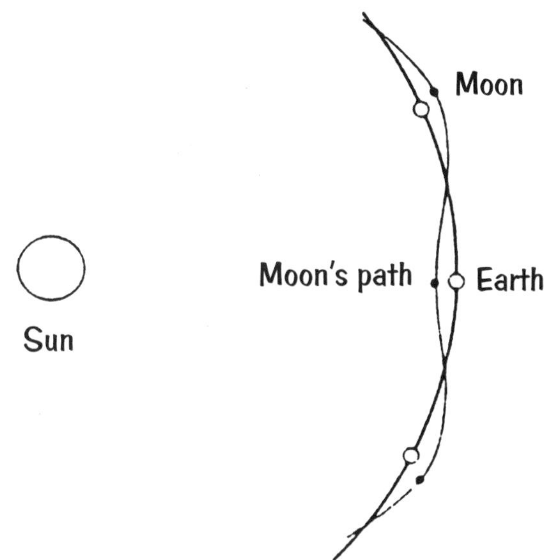

385. The Full Moon

Although the lighted area of the full Moon is only twice as large as that of the Moon at first or last quarter, the full Moon is about nine times brighter. Why?

386. The Moon Illusion—Luna Mendex

Few sights are more striking than the full Moon hanging low over the horizon. The Moon looks much bigger than when it is high in the sky. The effect cannot be due to atmospheric conditions, since in photographs the Moon's image is essentially the same size, 0.5 arc

The greatest number of eclipses, both solar and lunar, possible in one year is seven. The least number of eclipses possible in one year is two, both of which must be solar, as in 1984.

The lowest rising full Moon is always nearest the first day of summer. Hence it appears larger and more colorful ("spoon under the full honey Moon of June").

The ratio of the effect of the Moon on ocean tides to the effect of the Sun is approximately 7/3. One can show that 7/3 must also be the ratio of the mean densities of the Moon and the Sun, 3.34 g.cm^{-3}/1.41 g.xm^{-3}.

Only one month has ever elapsed without a full Moon, February 1866, an event that will not repeat itself for 2.5 million years.

All this talk about space travel is utter bilge, really.

—Richard Wooley (British astronomer royal, 1956)

Earth is the densest planet, at 5.52 g/cm^3, so it must have a lot of iron. In contrast, the Moon is much less dense, only 3.34 g/cm^3, and must be mostly rock. LIke Earth, and unlike the Moon, Mercury has a high density, 5.44 g/cm^3, so it must have plenty of iron. In fact, Mercury has far more iron than Earth. Mercury's "uncompressed" density, found by calculating the volume without the effect of the planet's weight, is higher than that of Earth or any other planet, 5.3 g/cm^3, compared to 4.4 g/cm^3 for Earth. Mercury must have an enormous iron core, almost as big as the planet itself.

In its crescent phase Venus is sometimes so bright it may be visible in the middle of the day and may even cast a shadow.

At any given location on Earth, a total solar eclipse happens once every 360 years.

degrees. In fact, the Moon is actually slightly closer by half the diameter of the Earth when it is overhead!

Another explanation, sometimes advanced, is that the eye is tricked into comparing the horizon Moon to the nearby objects—buildings, trees, hills, etc. However, this cannot be right either, because the illusion can be obtained over water or desert, where there are no familiar terrestrial objects for comparison. What is the correct explanation?

387. Setting Constellations

The Moon and the Sun appear larger when they are close to the horizon. Does the same effect happen with stars? In other words, do the constellations "expand" as they approach the horizon?

388. The Moon Upside Down?

Do the people in the Southern Hemisphere see the Moon upside down?

389. How High the Moon?

In winter, the Sun is low in the sky. If the Moon and the planets are near the ecliptic, the apparent path of the Sun in the heavens, why doesn't the Moon appear low in the sky, too?

390. "Earthrise" on the Moon?

Can you see the "Earthrise" or "Earthset" on the Moon?

391. Visibility of Mercury and Venus

Why are Mercury and Venus generally invisible at night?

392. Density of Earth

The mean density of the earth is 5.52 g/cm^3—that is, 5.52 times the density of water. In contrast, the four giant planets of the solar system have much lower densities: Neptune, 1.64 g/cm^3; Jupiter, 1.33 g/cm^3; Uranus, 1.29 g/cm^3; Saturn, 0.69 g/cm^3. Saturn could float in water! What is the reason for this contrast to the earth's density?

393. Rising in the West?

Are there any natural objects in the Solar System that rise in the west and set in the east as seen by observers on different planets?

394. Taller Mountains on Mars

The tallest mountain on Earth is not Mount Everest, but the Hawaiian volcano Mauna Kea, which rises 33,400 feet from the ocean floor, surpassing the height of Mount Everest by more than 4,000 feet. But only the top 13,796 feet of Mauna Kea show above the surface. Surprisingly, however, the tallest mountain on Mars, the volcanic cone Olympus Mons, is at least 80,000 feet high, and its base is 350 miles in diameter. Mars is only about half the Earth's size, yet some of its mountains are much taller than ours. Any explanation?

Venus receives 1.9 times more solar radiation than the earth, but Venus's sulfuric acid clouds reflect about 80 percent of that sunlight, so that Venus actually absorbs significantly less solar energy than the earth. Without the carbon dioxide that causes the greenhouse effect, Venus would be colder than the earth and only slightly warmer than Mars.

The present inhabitation of Mars by a race superior to ours is very probable.

—CAMILLE FLAMMARION (FOUNDER OF THE FRENCH ASTRONOMICAL SOCIETY, 1892)

The shortest known rotation period for an asteroid is displayed by 1566 Icarus (2 hours, 16 minutes). The slowest spinner is 280 Glauke (1,500 hours).

I am sorry to say that there is too much point to the wisecrack that life is extinct on other planets because their scientists were more advanced than ours.

—JOHN F. KENNEDY

Through space the universe grasps me and swallows me up like a speck; through thought I grasp it.

—Pascal Pensées

The first prize offered for communicating with extraterrestrial beings was the Guzman Prize, announced in Paris on December 17, 1900. The sum involved was 100,000 francs—but Mars was excluded, because it was felt that contact with the Martians would be too easy!

—Patrick Moore

Even though the Moon is much smaller than the Earth (about one-quarter the radius of the Earth), it takes 108 minutes to circle the Moon in the lowest orbit compared to 84 minutes for the lowest Earth orbit. The reason? The surface gravity on the Moon is only about one-sixth that on Earth.

*395. Going to Mars by Way of Venus!

The U.S. *Mariner* and *Viking* probes to Mars used the standard Hohmann transfer orbit, which minimizes the expenditure of fuel and follows an ellipse cotangent to the orbits of both Earth and Mars. The trip takes 7½ months. However, to get back to Earth, the expedition must wait for 1 year and 4 months before Earth and Mars are again lined up to make the return trip in a similar manner. Round-trip time—more than 2½ years! Surprisingly, a quicker way to get to Mars is to go by way of Venus. How is this feat possible?

*396. Where Are You?

Suppose you are in a windowless room aboard a wheel-shaped space station. The station is spinning about its hub to maintain normal simulated gravity. What simple test can you make to convince yourself that you are aboard a space station and not on Earth?

*397. Was Galileo Right?

One of the fundamental breakthroughs in physics, we are told, came when Galileo found that, neglecting air resistance, all bodies fall with the same acceleration. But is this result really independent of the mass of the falling object? What if the object is as large as, say, a massive asteroid?

Answers

Chapter 1 Temperature Risin'

1. Thermos Delight!

Pour one-half of the cold water in thermos *B* into container *D* and insert *D* into thermos *A*. The final temperature of the waters in both *A* and *D* will be 60 °C. Now pour the 60 °C water in container *D* into thermos *C*. Repeat the procedure with the other one-half of the cold water in thermos *B*, again using container *D* and thermos *A*. The final temperature in container *D* and thermos A will be about 47 °C. Now pour the water in *D* into thermos *C*, and the final temperature of the 1 liter of water in thermos C will be about 53 °C. The water in thermos *A* will be about 47 °C.

2. Boiling Water with Boiling Water

No. Pure water boils at 100 °C. When the water in the small container reaches 100 °C, no thermal energy will be transferred from the boiling water to the 100 °C water inside the small container. To convert water to vapor at 100 °C takes an additional 580 calories per gram. Therefore, no boiling is achieved.

3. Gas and Vapor

Yes. A vapor is a gas below its critical temperature. For water the critical temperature is 374 °C. Above this critical temperature, water vapor will not condense into droplets, no matter how much pressure is applied.

The word "steam" is used often in reference to both the invisible water vapor and the visible mist of water droplets. Steam is defined as water vapor at or above the boiling point of water—that is, 100 °C at normal pressure. In daily experience, steam is what's seen around the spout of a teapot!

4. Ice in Boiling Water?

The water at the bottom of the tube remains cold enough so the ice melts very, very slowly. The hotter water at the top is less dense and remains at the top. The poor conductivity of the water limits the thermal energy transfer rate to the ice at the bottom, so the important physical factors favor the ice.

Iona, M. "Another View by Iona." Physics Teacher *28 (1990): 444–445.*

5. Two Mercury Droplets

Assume the ideal case of no thermal energy transfer from the droplets to the surroundings. We can determine that the surface area of the new droplet is less than the total surface area of the original two droplets. The decrease in the surface area means a decrease in the energy of surface tension necessary to pull the mercury into its shape. The extra energy raises the temperature of the final droplet. If this process occurs on a level, flat surface, such as a glass plate, then one must also consider the gravitational potential energy changes, and an even greater temperature change is possible.

6. Drinking Bird

The familiar drinking bird derives its energy from the temperature difference between its body and its head. The bulb is at room temperature, but the head is cooler due to evaporation of water from the large surface area of the felt on the outside of its head and beak. With this temperature difference, the pressure of the vapor in the body part is greater than the pressure in the head, so some of the methylene chloride is pushed up the tube, shifting the center of gravity so the head goes down and the beak becomes immersed in the water. In that position, the lower end of the tube is open to the vapor, and liquid drains back into the bulb at the bottom of the body. The bird tips upright, and the cycle starts anew.

Bachhuber, C. "Energy from the Evaporation of Water." American Journal of Physics *51 (1983): 259–265.*

Crane, H. R. "What Does the Drinking Bird Know about Jet Lag?" Physics Teacher *27 (1989): 470.*

Mentzer, R. "The Drinking Bird: The Little Heat Engine That Could." Physics Teacher *31 (1993): 126.*

7. Room Heating

Paradoxically, the answer is that the total energy of the air in the room remains the same. When the air temperature is increased by the heater, the air in the room expands and leaks some air to the outside through pores and cracks in the walls. This leaking air carries with it the energy added by heating!

For air acting as an ideal gas, the energy content of the air in the room is independent of the temperature when the pressure must remain fixed. From the relation $PV = nRT$ you know that the increase in volume V is directly proportional to the temperature T increase when the pressure P is held constant. R is the gas constant and n is the number of moles of ideal gas. Imagine a room allowed to expand into its new, larger volume with the same number of moles n. Now cut out of this larger volume a room with the original volume, and you have decreased n by the same factor. Therefore the ideal gas relation tells us that the total energy is the same as before.

8. Shivering at Room Temperature

Shivering is not the normal body response, for not much transfer of thermal energy per second occurs. For simplicity, we ignore the thermal effects associated with wearing clothing. At least three effects must be considered:

1. The surface of the skin is at a much lower temperature than the 37 °C internal body temperature.
2. The air is a poor thermal conductor, and without convection currents, thermal energy transfer by air conductivity is inefficient.
3. Evaporation of water from the skin surface depends upon the air speed nearby. If the air is quiet, a stagnant warm-air layer forms over the skin, and the evaporation rate is small. But a gentle 3 mph breeze doubles the wind-chill effect when compared to air moving at less than 1 mph!

9. Identical Spheres Are Heated

No. The suspended sphere will be warmer. The gravitational potential energy changes of the spheres will be different as they expand. Some of the thermal energy goes into raising the center of gravity of the sphere on the table, so its temperature rise is less than expected. The expansion of the hanging sphere lowers its center of gravity, so its temperature rises more than expected.

10. Cooking Hamburgers

Hamburgers do cook faster when the outside is not charred, which usually happens when cooked over a high flame on the barbeque. The charred meat on the outside is a poor thermal conductor, so the inside meat takes longer to reach the required temperature. Experience teaches one this rule of thumb: Hamburgers cooked slowly cook faster.

11. Cooking Hamburgers versus Steaks

On a slab of steak, most of the surface bacteria will be on the outside area and not inside, and they will be killed quickly when the steak is heated. With ground beef, the surface bacteria are dispersed throughout the hamburger patty, so the hamburger should be cooked thoroughly to destroy these bacteria.

12. Gasoline Mileage

A gallon of cold gasoline results in greater mileage because it contains more molecules. Like most substances, gasoline expands when the temperature increases. And if the measuring container does not expand also to compensate exactly, a gallon of warm gasoline doesn't go as far.

13. Triple Point of Water

At 273.16 K, all three phases of water—solid, liquid, and gas—coexist in equilibrium in a sealed vessel with no other substance present. The saturated vapor produces the pressure. If a little extra thermal energy is gained from or lost to the surroundings, the temperature will remain the same. If some energy enters the system, some ice will melt to decrease the volume of the liquid plus solid phases slightly, but a little more evaporation will occur to maintain a constant pressure.

14. Cold Salt Mixtures

Most freezing mixtures of salt and ice employ the very same materials to supply the thermal energy for melting itself. First, adding salt to water lowers the water's freezing point because the salt molecules (and ions) come between the water molecules to hinder their bonding attempts. Some ice present will therefore melt immediately, a physical change that requires 80 calories per gram of ice. This energy will be transferred from the unmelted ice and the water nearby. Percolation effects ensure good mixing so that even the last granules of ice will melt to make a very cold salt solution.

On the microscopic scale, some original translation energy of water molecules has increased the electrical potential energy of the ice molecules. Since the original random kinetic energy of water molecules is now shared among the original water molecules and those initially bound in the ice, the average random kinetic energy per molecule is lower (i.e., the temperature of the mixture is lower).

15. To Warm, or Not to Warm?

By blowing gently, you bring 37 °C warm carbon dioxide from your lungs to heat the cooler skin of your hands. By blowing harder, two effects result: (1) cooler room air is pushed into the stream by the Bernoulli effect, and (2) there is more evaporation per second from the skin, requiring thermal energy from the surface. Both effects produce a sensation of coolness.

16. Modern Airplane Air Conditioning

Taking in fresh air from outside at 30,000 feet altitude is costly energy-wise. The fresh air must first be compressed to about 1 atmosphere, which raises its temperature significantly above normal cabin temperatures, then cooled to the appropriate temperature. Both processes require significant energy from the fuel, which could be used to fly the plane farther. Therefore, fuel savings can be achieved by recirculating a greater percentage of the air and bringing in less fresh air per mile of travel. Some people claim that this increased recirculation of room temperature air also recirculates more bacteria, which could be a health problem.

17. Out! Out! Brief Candle

The flame goes out and the water level in the glass rises. As the flame burns, the gas inside the glass is warmed and expands. Some of the gas bubbles out under the mouth of the glass. (A careful look at the bubbling will verify this process.) When the flame decreases from oxygen deprivation, the remaining trapped gas cools, its pressure decreases, and the ambient atmosphere pushes more water into the glass. Eventually there is no more available oxygen to burn, so the flame goes out.

Most people make the mistake of thinking that the oxygen molecules burning with the evaporating candle wax hydrocarbon molecules reduces the number of molecules in the gas above the liquid. But this is not the case. There would be more molecules produced as products of combustion than reacted initially. Just look at the balanced chemical equation. For example

$$2\ C_6H_{14} + 19\ O_2 \rightarrow$$
$$12\ CO_2 + 14\ HOH$$

where 21 initial molecules produce 26 product molecules.

18. Piston in a Beaker

The enclosed space above the liquid surface contains its saturated vapor. If the piston is raised slowly, the space is filled to atmospheric pressure by the saturated water vapor. Therefore the water level in the tube does not change.

If the piston is raised quickly, the vapor pressure in the tube will be less than atmospheric because the water vapor will not form fast enough. So the water rises to a height such that the atmospheric pressure is equaled by the sum of the hydrostatic pressure and the saturated vapor pressure. Eventually the pressure of the water vapor will push the water back down. With boiling water, the vapor pressure will remain constant with both a rapid and a slow rise of the piston.

19. Milk in the Coffee

The experimental results reveal that black coffee cools faster than white coffee under the same conditions, by as much as 20 percent. Any draft of air can have an enormous effect on the cooling rate, so comparisons must be done in quiet air under the same insulating conditions. The cooling time is then approximately proportional to the ratio of the volume to the total surface area of the liquid, other factors being equal. Newton's law of cooling states that the cooling rate is proportional to the temperature difference between the outside surface of the coffee cup and the ambient air. This law tends to hold very well.

Under most household conditions, one should go ahead and add the milk first if the wait is to be fewer than ten minutes or so. Although the slopes of the cooling curves are different, they do not cross because the temperature decreases exponentially.

Rees, W. G., and C. Viney. "On Cooling Tea and Coffee." American Journal of Physics *56 (1988): 434–437.*

20. Energy Mystery

Half of the initial potential energy was converted into thermal energy by the internal friction and the friction against the walls. Without the friction, the liquid would oscillate between the two containers forever.

21. Dehumidifying

When warm air is cooled, tiny water droplets will form from the water vapor. There will be more low-speed collisions between water molecules at a lower temperature, so more coalesce into droplets. The cool humid air is not as comfortable as cool dryer air, so dehumidification is necessary.

22. Refrigerator Cooling

Initially, the cooler air in the refrigerator does cool the room air a little bit, depending upon the relative volumes, the mixing, and the temperature difference. However, when the refrigerator starts up again, more thermal energy will be released into the room by the cooling coils in the back than is absorbed by the cool air emanating from the refrigerator front, as dictated by the second law of thermodynamics. The room will become warmer.

23. Air and Water

Although both quiet air and still water are poor thermal conductors, the water is still a much better thermal conductor than air. The higher rate of thermal energy "flowing" from your body into the swimming pool water makes the water feel colder.

24. Hot and Cold Water Cooling

Under certain conditions the hot water will cool faster than the cold water and begin to freeze first!

First, notice that the pails do not have lids, and recall that wood is a very poor thermal conductor. The following argument works well for wooden pails but not so well for pails that are good thermal conductors.

The main cooling effect is rapid evaporation from the top surface of the hot water, followed by significant mixing of the hot and cooler water from top to bottom. The evaporation plus convection produce a rapid rate of thermal energy transfer to the surroundings if the starting temperature is high enough. For these wooden pails, the thermal energy transfer rate is many times the transfer rate by conduction through the wooden walls of the pails. Moreover, up to about 26 percent of the water in the original hot water pail might be evaporated away, leaving much less water to freeze.

As stated, the mass loss in cooling by evaporation is significant. For an extreme example, water cooling from 100 °C to 0 °C will lose 16 percent of its mass, and another 12 percent of the mass will be lost on freezing. The total mass loss is therefore $16\% + 12\% \times (100 - 16) = 26\%$.

This paradoxical fast cooling of hot water was reported by Francis Bacon in *Novum Organum* (1620). In places that experience long winters, such as Canada and the Scandinavian countries, it has become part of everyday folklore. For example, it is believed that a car should not be washed with hot water because hot water will freeze on the car faster than cold water, and that a skating rink should be flooded with hot water because it will freeze more quickly.

Auerbach, D. "Supercooling and the Mpemba Effect: When Hot Water Freezes Quicker Than Cold." American Journal of Physics *63 (1995): 882–885.*

Chalmers, B. "How Water Freezes." Scientific American *238 (1959): 114–122.*

25. Ice-skating on a Very Cold Day

The static friction coefficient is much greater as the ice surface becomes colder. Therefore, the maximum value of the static friction will be significantly greater also, so gliding becomes very difficult.

Note: The ice surface near 0 °C *always has a very thin film of water,* which acts as a lubricant between the ice surface and the ice skates. In fact, all simple solids have a thin layer of liquid on their surface, even well below their bulk melting points, because the free energy of the surface is reduced when a thin surface layer is in the liquid phase.

Also note that there is no experimental verification that the pressure from the small contact area of the ice skate runner is great enough to cause a melting of some ice at the surface. It is known that a pressure of about 140 atmospheres would be required for bulk melting, much more than you get with sharp skates!

Wettlaufer, J. S., and J. G. Dash. "Melting Below Zero." Scientific American *282 (2000): 50–53.*

White, J. D. "The Role of Surface Melting in Ice Skating." Physics Teacher *30 (1992): 495–497.*

26. Singing Snow

At air temperatures near 0 °C, a very thin film of water on each ice crystal lubricates the rubbing between them when the shoe pushes on them. At much lower temperatures there is no water film on the ice crystals, so the friction between them in response to the shoe pressure produces a relaxation oscillation called a "squeak."

27. Contacting All Ice Cubes!

Ice cubes in a bucket contact each other in small areas. Originally, each ice cube has a very thin film of water on its surface, but in the contact area the surface exposed to air exists no longer. So a little bit of thermal energy is removed from the water, freezing occurs, and the ice cubes stick together, a process called "ice sintering." Essentially, the free energy is adjusting on the surface and in the bulk solid.

28. Hot Ice

Yes. Ice at 20,000 atmospheres melts at 76 °C, hot enough to burn skin!

29. Walden Pond in Winter

Fish and all living organisms can be thankful that water expands from about 4 °C down to 0 °C. Otherwise, all life might have died out during one of the ancient ice ages.

Here's the argument: Start at 6 °C for both air and water temperature and slowly lower the air temperature above the water. At 5 °C the 5 °C water at the surface is more dense than the 6 °C water below, so mixing occurs, bringing up warmer water to cool at the surface and sending down cooler water. At 4 °C the surface water at 4 °C is still more dense, so mixing continues and the water at lower depth continues to cool to 4 °C.

But at 3 °C the surface water is less dense, so this 3 °C water stays at the surface and no more mixing occurs. That means the water at lower depths does not go much below 4 °C because it can no longer cool efficiently. Cooling occurs now by conduction only, a very poor process compared with the convection currents before.

And when the ice forms on the surface, its thermal conductivity is even worse than water's, so the ice acts as a thermal insulator between the water and the cold air. The water beneath the ice does not freeze, and life goes on.

30. Lights Off?

During winter there is no energy advantage to turning off unnecessary incandescent lights. In the summer, every extra light adds thermal energy to the room that must be removed by the air conditioning, so one should turn off the lights.

Incandescent lights are very efficient heaters, and even the emitted light (about 10% of the energy) will eventually be converted to thermal energy upon absorption by the walls, the furniture, and other objects.

In winter, the thermal energy no longer supplied by the extinguished incandescent bulb must be supplied by the heating system, which is usually not as efficient as the electricity generation and transmission. However, having the bulb on may cost a bit more money because electricity is often a more expensive method for heating a building. Also, lightbulbs cost money to replace.

P. A. Bender. "Lights as Heaters." Physics Teacher *13 (1975): 69.*

31. The Metal Teakettle

No, if the metal handle is stainless steel or any other material that is a poor conductor of thermal energy. Some types of stainless steel are extremely poor thermal conductors.

32. Frozen Laundry

The ice is subliming from the solid to the gas phase without becoming liquid.

33. Ice Cream in Milk

The tongue and the walls of the mouth are sensing the rate of transfer of thermal energy from the living tissue to the ice cream mixture. If the ice cream were mostly ground-up ice crystals, then adding milk would increase the contact area immensely, and you would experience more thermal energy transferred per second. The combination would feel colder. In addition, the liquid is a much better thermal conductor than the ice crystals, which have trapped quiet air (i.e., no convection currents), so the combination would feel colder. Both effects contribute to the sensation of coldness.

34. Wearing a Hat in Winter

Up to 30 percent of body cooling can be from the head. Wearing a hat could reduce this cooling very effectively to help keep the body warm. Incidentally, Aristotle thought that the head was the great cooling agent for the body.

35. Car Parked Outside

On a clear night, the roof of the car "sees" the night sky of the universe, which has a temperature of about 285 K, so the roof radiates away an enormous amount of energy per second and cools. Moisture in the air condenses on the cool roof, which is wet by the morning.

On a cloudy night, the roof cannot "see" the night sky. Instead, the roof "sees" the clouds, which are warmer than 0 °C (about 300 K). So the roof maintains about the same temperature as the ambient air, and no moisture forms.

36. Two Painted Cans of Hot Water

All factors being identical except for their colors, both cans should cool at the same rate. Just because one can is black and the other is white in the visible part of the electromagnetic spectrum does not mean that they are different in the infrared (IR). Their IR characteristics, not their visible light characteristics, determine the cooling rate by radiation.

Bartels, R. A. "Do Darker Objects Really Cool Faster?" American Journal of Physics *58 (1990): 244–248.*

Ristinen, R. A. "Some Elementary Energy Questions and (Wrong) Answers." American Journal of Physics *50 (1982): 466–467.*

37. Sunshine

At least two factors determine the ambient air temperature in the first few meters above the ground: the ground temperature and the amount of direct solar energy. In winter the ground is already cool, so warmer air currents passing near the ground will become cooler. In winter the sunlight enters at an angle that is less than ninety degrees to the ground surface, so less energy is delivered to warm the ground than in summer. Both effects tend to keep the ambient air cool. Wind-chill and other effects also occur. *Note:* Contrary to intuition, the sunlight does very little heating of the air by direct absorption.

38. Physicist's Fireplace

Yes. Instead of having the fire burn between the logs, one should have the logs supported so that one can see the hottest glowing region from the room. This configuration usually requires the front log to be removed to leave an opening, while the upper logs need to rest on supports. Then much more infrared radiation will be emitted into the room for heating.

Walker, J. "...On Making the Most of a Fireplace." Scientific American *257 (1978): 140–148.*

39. Blackbody Radiation

The microwave background radiation in the universe corresponds to a temperature of 2.8 K and shows no absorption lines. The radiation from an oven is distorted by the absorption lines of the atoms in the material of the oven.

*40. Uniqueness of Water

Water expands for the last few degrees above its freezing temperature when cooled. By the way, water expands about 11 percent in going from liquid at 0 °C to ice at 0 °C, enough to burst most containment vessels, including iron water pipes.

*41. Blowing Hot and Cold

The Ranque-Hilsch vortex tube can separate air into a hot air stream and a cold air stream without any moving parts because the air initially cools by expansion upon entry. Near the inlet there is a vortex with greater speeds near the tube axis and slower speeds nearer the tube wall. The air moving toward the hot end of the tube experiences viscous interactions between the warmer air near the axis and the cooler air, resulting in work being done to heat the outer regions of the air as it exits the hot end of the tube. The core of the vortex expands as it moves toward the cool end and exits.

Chapter 2
Color My World

42. Corner Mirrors

Your image *in the corner* shows no change in handedness with the perpendicular corner mirrors, in contrast to the reversed images seen in each of the single plane mirrors. This result is a consequence of the image being both a left-right reversal and a front-back reversal.

Galili, I.; F. Goldberg; and S. Bendall. "Some Reflections on Plane Mirrors and Images." Physics Teacher *29 (1991): 471.*

43. The Vanishing Elephant

The elephant actually stays in the cage. When the time for disappearance comes, two large mirrors are slid quickly into place and the audience then sees the side walls of the stage. These side walls are designed so that the reflected light from them matches the backdrop of the stage, with no elephant visible. The two large plane mirrors are at right angles to each other, with the line of contact forward, toward the audience. A strobe light is used to conceal the brief motion of the mirrors. Then the elephant is quickly led out through a door unseen by the audience.

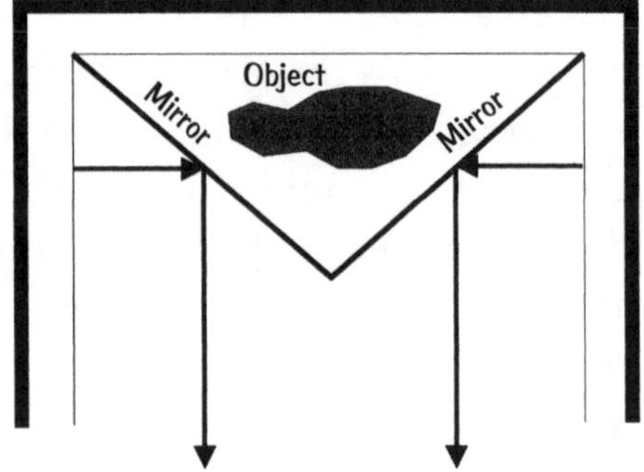

Edge, R. D., and E. R. Jones Jr. "Optical Illusions." Physics Teacher *22 (1984): 591–593.*

Ruiz, M. J., and T. L. Robinson. "Illusions with Plane Mirrors." Physics Teacher *25 (1987): 206–212.*

44. Floating Image

The real image is produced by two reflections, one from each concave mirror's inside surface, before the light ray exits. An upright object on the bottom appears as an upright real image as determined by looking at the image and by ray tracing.

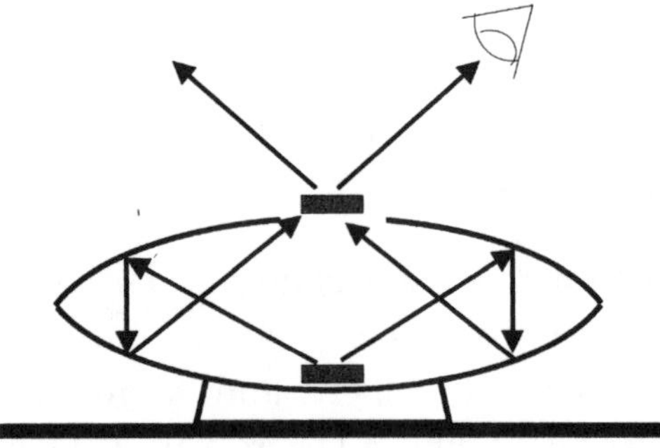

Sieradzan, A. "Teaching Geometrical Optics with the 'Optical Mirage.'" Physics Teacher *28 (1990): 534–536.*

45. Lighting an Image?

The real image will be appropriately illuminated in exactly the region where the light is aimed. One can trace light rays from the flashlight going through the real image back to strike the real object on the bottom mirror. Therefore the original object is illuminated by the flashlight and so is the real image.

Mackay, R. S. "Shine a flashlight on an Image." American Journal of Physics *46 (1978): 297.*

46. Laser Communicator

You should aim the laser directly along the line of sight to the space station. There will be a very slight angle difference for red light and for blue light, but the space station receptor would be large enough compared to the beam diameter and to the separation distance that it wouldn't make a difference.

Hewitt, P. "Figuring Physics." Physics Teacher *28 (1990): 192.*

47. Bent Stick

The contradiction is only apparent. The eye of the observer receives reflected light from the bottom of stick *B*. But the light ray from *B* changes direction at the water-air interface *C* following the path *BCD* to reach the eye. To the observer, the light appears to have come straight from behind *C*, or from a point around *E*. Notice that point *E* is higher than *B*, so the stick appears to be bent upward.

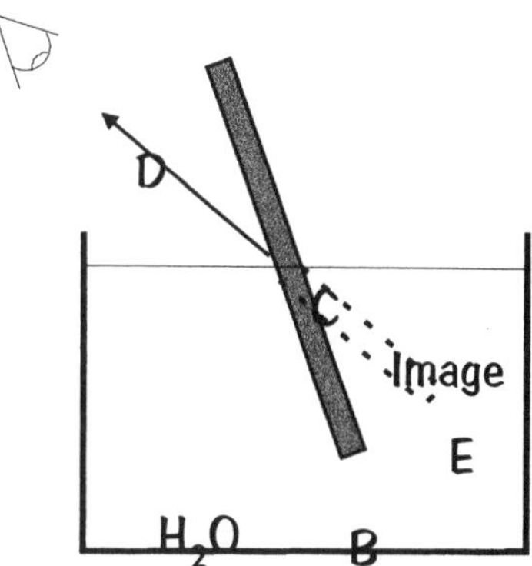

48. The Pinhole

Yes. The geometry of similar triangles reveals that the ratio of the sun's diameter to the sun's image diameter equals the ratio of the distance to the sun divided by the image distance from the pinhole. One knows the other three quantities, so the sun's diameter can be determined.

Young, M. "Pinhole Imagery." American Journal of Physics *40 (1972): 715–720.*

———. *"Imaging without Lenses or Mirrors."* Physics Teacher *27 (1989): 648.*

49. Window

The open window appears black or very dark in the daytime because most of the light enters the opening and does not exit. This same behavior explains why the pupil of your eye is black. In fact, even the black print on this page absorbs most of the incident light. The information you read in these words is actually determined by the reflections of light from the white paper surrounding the black letters!

50. Window Film

No, the window film helps in winter, too. The film also transmits less infrared radiation. So in the winter, more infrared energy stays inside the room.

Wald, M. L. "Windows That Know When to Let Light In." New York Times *(August 16, 1992), p. F9.*

51. Rainbow

There are two ways Nature resolves the problem. One, the raindrops are not actually spherical, so identical geometrical conditions do not occur at each air-water-scattering interface. Two, some light always emerges through the interface even for total internal reflection.

52. An Optical Puzzle

The image will flip over and turn right side up—that is, the 90-degree rotation of the mirror results in a 180-degree rotation of the image. A ray trace diagram would show why this behavior is expected.

Derman, S. "An Optical Puzzle That Will Make Your Head Spin." Physics Teacher *19 (1981): 395.*

Holzberlein, T. M. "How to Become Dizzy with Derman's Optical Puzzle." Physics Teacher *20 (1982): 401–402.*

Wack, P. E. "Cylindrical Mirrors." Physics Teacher *19 (1981): 581.*

53. Rearview Mirror

The rearview mirror is wedge-shaped, with a silvered surface in back. The wedge angle is between three and five degrees. During the daytime one sees the reflection off the back surface. At night, after the mirror has been tilted, one sees the poorer reflection off the front surface, which is not silvered. There is still a reflection off the silvered surface, but this reflected light misses the eyes.

Jones, E. R., and R. D. Edge. "Optics of the Rear-View Mirror: A Laboratory Experiment." Physics Teacher *24 (1986): 221.*

54. Colors

False. Most of the time the blouse looks green because the combination of colors selectively scattered to our eyes makes some shade of green. Surprisingly, there usually is no green light of the spectrum going into our eyes. Our eye-brain system fools us repeatedly in looking at colors, but a spectrometer will reveal the true colors—the actual frequencies of the light—scattered by the blouse.

55. Primary Colors

True and false. Red, green, and blue are not the only primary colors for light. Any three colors of light can be used as primary colors as long as they are orthogonal—that is, the third color cannot be made from a combination of the other two.

Often a second condition is invoked by requiring the triplet of primary colors to produce the widest possible range or variety of colors. This condition is somewhat subjective, however, because each person has a slightly different response to light. Consequently, no one lists three specific light frequencies as the best triplet of primary colors.

Feynman, R. P.; R. B. Leighton; and M. Sands. The Feynman Lectures on Physics. Vol 1. *Reading, Mass.: Addison-Wesley Publishing, 1963, page 35–6.*

56. Diamond Brilliance

You would see elliptical patches of colored light with a rainbow of colors. These colors appear because the blues separate from the reds a little more at each of the four interfaces along the path. The index of refraction is slightly different across the frequencies of the visible spectrum.

Friedman, H. "Demonstrations of the Optical Properties of Diamonds." Physics Teacher *19 (1981): 250–252.*

57. White Light Recombined

A converging lens works best if placed close to the prism that initially separated the spectrum of light. A flat paper sheet for the image can then be moved the required distance to see the white light recombined from the colored light rays.

MacAdam, D. L. "Newton's Theory of Color." Physics Today *38 (1985): 11–14.*

Pregger, F. T. "Recombination of Spectral Colors." Physics Teacher *20 (1982): 403.*

58. Prisms

No. Two prisms cannot recombine the spectrum of light rays from a prism back into the original narrow beam of white light. This common fallacy is perpetuated to this day in many texts. Empirically, one gets parallel rays of color exiting in a wide colorful beam, not a single narrow beam of white light. Four identical prisms are required to recombine the rays into white light.

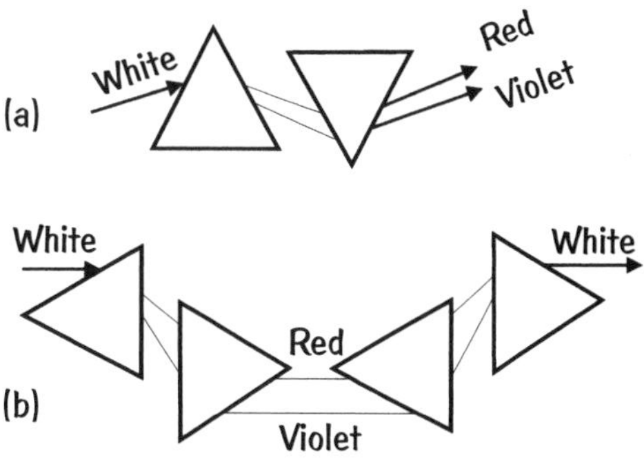

59. Squinting

The squinting works because the eye, like a pinhole camera, will have an infinite depth of field—that is, the image will be in focus over a wide range of distance, so squinting will work for improving both nearsightedness and farsightedness. Essentially, for nearsightedness the rays entering the eye farther off the central optical axis do not focus on the retina, so blocking these rays will improve the image.

Pinhole glasses are not a very good solution because the field of vision is very narrow. The pinhole is far from the eyeball, and the light intensity is drastically reduced.

However, one can improve myopic vision by using a brighter light, the intensity allowing the pupil to reduce its diameter to block the rays far off the optical axis.

Keating, M. P. "Reading through Pinholes: A Closer Look." American Journal of Physics *47 (1979): 889–891.*

Mathur, S. S., and R. D. Bahuguna. "Reading with the Relaxed Eye." American Journal of Physics *45 (1977): 1097–1098.*

60. Polarized Sunglasses

Through the left eyeglass, objects would appear normal, with less glare from the horizontal surfaces. But through the misaligned right eyeglass, many of the distant objects would appear to be *extremely close* and somewhat dim. The brain has interpreted the distant objects to be very close to the observer, on the right side of the nose, because the objects are so bright!

By the way, polarized stereoscopic glasses for movies or slides have polarization directions perpendicular to each other, often at ± 45 degrees. Other types employ "shutters" that expose one eye or the other to the light from the screen at the proper times.

Hodges, L. "Polarized Sunglasses and Stereopsis." American Journal of Physics *52 (1984): 855.*

61. Visual Acuity

The human eye has three types of cone sensors for visible light: red, green, and blue. If the area density of blue cones is equal to or less than that for the green cones, then the angular resolution will be determined by this area density instead of the wavelength criterion. The human eye does not have a great enough area density of blue cones.

The visual acuity is also dependent upon the light intensity because the pupil reduces its diameter in brighter light, eliminating those rays entering farther from the optical axis.

Kruglak, H. "Another Look at Visual Acuity." Physics Teacher *19 (1981): 552–554.*

62. Laser Speckle

The granular pattern seen in the reflected laser light is called "speckle," an effect caused by interference between different rays of the laser light that are scattered to your eye and traveling different distances. If you move your head and see the speckle pattern move in the opposite direction, you are nearsighted. You are focusing your eyes in front of the real surface, which then appears to move opposite your head movement. Farsighted people experience the spot moving in the same direction as the head movement.

Caristen, J. L. "Laser Speckle." Physics Teacher *25 (1987): 175–176.*

Walker, J. "The 'Speckle' on a Surface Lit by Laser Light Can Be Seen with Other Kinds of Illumination." Scientific American *261 (1982): 162–169.*

63. The Red Filter

You do not see the red *R* because the white paper reflects about as much red light per unit area as the red crayon *R* does, and this red light passes through the red filter. But you do see a "shadow" of the blue *B*, because the blue crayon reflects very little red light and you see this dark contrast to the red-light intensity from the white paper.

Kernohan, J. C. "Red, White, Blue, and Black." Physics Teacher *29 (1991): 113.*

64. Red and Blue Images

No. The red and blue images of the same object are different sizes because the blue light is refracted by the eye through a slightly bigger angle. If the blue image is in focus on the retina, then the red image would focus slightly behind the retina, so the red image appears slightly larger and perhaps a little fuzzy.

65. Colors in Ambient Light

In most cases, the colors of your clothes whether viewed inside the room or outside in the sunlight appear to be the same, even though the ambient light is drastically different! The eye-brain system seems to subtract out the ambient light differences so that the colors appear to be nearly the same. The physiological mechanism that achieves this effect is still being investigated.

66. Seeing Around Corners?

Image formation is crucial to seeing but not to hearing. Hearing involves a detector system that is small compared to the wavelength of sound. Therefore hearing depends upon the temporal variation and not the shape of the wave fronts.

Seeing is more complex because it requires image formation. Images depend upon the phase relationships of adjacent light rays. Light rays diffracted or scattered into the geometrical shadow of a corner no longer have the original phase relationships. As they diffract around the corner, the light rays change their relative directions and phases. On the average, they cancel out, so the object would not be seen. In contrast, a flat mirror placed to reflect the light around the corner maintains both the parallel ray directions and the proper phases.

Ferguson, J. L. "Why Can We Hear but Not See around a Corner?" American Journal of Physics *54 (1986): 661–662.*

67. Stereoscopic Effect

The stereoscopic effect appears because each eye sees a slightly different pattern of sparkles of reflected light. The geometry of our binocular vision reveals just where in space each image should appear. Likewise for the milling marks on an aluminum sheet, many of which appear to be floating above the actual metal. Such effects can be quite startling. A stereogram from two pictures of the same scene taken at two slightly different angles produces 3-D images when viewed properly, and this effect is related to seeing the sparkles.

Hulbert, E. O. American Journal of Physics *15 (1947): 279.*

68. Eye Color

Eye color ranges from light blue to dark brown, but the color can vary from eye to eye and within the same iris. Eye color is determined solely by the concentration of identical pigment cells called melanocytes—all of which contain the same pigment, melanin, no matter what the eye color!

Light reflects off the back of the iris. The higher the concentration of melanocytes there, the greater the opacity and the deeper the color. Blue and green eye hues indicate sparse melanocyte distribution. Brown eyes require melanocytes to also be at the front of the iris. Newborns do not yet have melanocytes at the iris front, but in a few months some of them migrate there.

Albinos lack pigment altogether. The pink eyes are created by light reflecting off the blood vessels at the retina.

Yulsman, T., ed. "Eye Color: An Optical Illusion." Science Digest *91 (1983): 92.*

69. Metal Clothing

Most of the thermal energy from a red-hot slab of metal heats you via infrared radiation (IR), electromagnetic energy with a frequency just lower than the red part of the visible spectrum. Metals are excellent reflectors of electromagnetic radiation, so a metal coating provides an effective shield against the infrared radiation emitted by the hot slab.

The thermal blanket in emergency kits for backpackers utilizes the high infrared reflectivity of a thin metal film to reflect IR onto the human body. Wrapped around a person, this combination of metal film on a plastic sheet keeps the small layer of air between the body and the blanket pretty toasty warm even in freezing-cold ambient temperatures.

*70. The Sky Should Be Violet

The sky would be violet instead of blue if Rayleigh scattering were the only important factor here. But the sunlight reaching the atmosphere does not have the same intensity for all colors. Instead, the sunlight has a peak intensity in the greens corresponding to the 6,000 K temperature at the sun's surface. Therefore the total intensity of violet light in the solar spectrum is less than the total intensity of the blue light, and the latter dominates the Rayleigh scattering in the sky.

As an additional effect, our eyes are less sensitive to the violets than to the blues, so the dominance discussed above is enhanced by our physiological response.

*71. Crooke's Radiometer I

Crooke's radiometer was first investigated in 1874 and has been stirring up controversy ever since. What makes this device even more startling at first is that one can remove practically all of the vane surface inside the perimeter to leave just wirelike edges and the device works just as well as before! Experiments in the 1920s demonstrated that the radiometer force occurred at the edges of the vane.

With a very good vacuum inside, the vane turns with the silvered side moving away from the light source, as expected. But most radiometers contain air at low pressure, about 0.1 mm of Hg, so the air molecules must play a role in making the radiometer turn with the blackened side moving away from the light source. We need to consider the effects at the edges of the vane only, for the forces acting on the vast surface area on both sides cancel each other. The blackened side at the edge will be slightly warmer than the silvered side at the edge. Consider an air molecule approaching the blackened edge within the mean free path distance, about 0.6 millimeter. This molecule will collide with a molecule rebounding from the vane edge and with molecules passing the vane edge from the cooler side. But the latter are less energetic, and their ability to turn away the incoming air molecule is less efficient. Therefore the blackened edge will suffer more collisions per unit area per second than the silvered edge. This excess is responsible for driving the vanes in the reverse direction opposite the force from the photon momentum exchange effects.

Woodruff, A. E. "The Radiometer and How It Does Not Work." Physics Teacher *6 (1968): 358–363.*

*72. Crooke's Radiometer II

The vanes can be made to rotate backward by cooling the radiometer in a refrigerator or by first heating the radiometer to a temperature above room temperature and then letting it cool. Either procedure must make the blackened side slightly cooler than the silvered side, an opposite condition to the normal operation at room temperatures. (See the previous problem for more details.)

Bell, R. E. "The Reversing Radiometer." American Journal of Physics *51 (1983): 584.*

Crawford, F. S. "Running Crooke's Radiometer Backwards." American Journal of Physics *53 (1985): 1105.*

———. "Running Crooke's Radiometer Backwards." American Journal of Physics *54 (1986): 490.*

Woodruff, A. E. Physics Teacher *6 (1968): 358.*

*73. Fracto-Emission of Light

For the tape being pulled off the glass, electrically charged particles exist along the line of separation between the tape and the glass. When the tiny gap is first formed, the potential difference per centimeter across the gap is great enough to break down the air, and a spark jumps the gap. A similar effect occurs when breaking certain types of candy wafers.

Walker, J. "How to Capture on Film the Faint Glow Emitted When Sticky Tape Is Peeled off a Surface." Scientific American *266 (1987): 138–141.*

*74. Perfect Mirror Reflection

Alternating layers of metallic and dielectric materials can produce a mirror capable of reflecting nearly all the incident light at any angle at any frequency! Combining the best properties of the metallic and the dielectric mirrors, this feat of making a perfect mirror was first accomplished in 1998.

Fink, Y., et al. "A Dielectric Omnidirectional Reflector." Science *282 (1998): 1679–1682.*

Chapter 3
Splish! Splash!

75. Air Has Weight!

About 2.2 pounds per cubic meter, or about 1 kilogram, so answer (f). Very few people guess a value greater than 2 ounces!! Knowledge of one way to quickly and accurately estimate a value of about 1 kilogram develops from knowing that 22.4 liters of air have a mass of about 28 grams and that 1 cubic meter contains 1,000 liters.

76. Damp Air

The damp air weighs less because water molecules of molecular weight 18 have replaced more massive molecules, for example, nitrogen, of molecular weight 28, and oxygen, of molecular weight 32. Both volumes have the same numbers of molecules. Therefore, a humidity increase lowers the barometer reading. The barometer pressure value gets lower as a storm approaches.

77. The Pound of Feathers

The pound of feathers weighs more. The larger buoyant force of the air acting on the larger volume of a pound of feathers (as the scale weighs them in air) must be compensated by more than a pound of feathers (as weighed in a vacuum) to achieve the same scale reading of one pound. Phew!

Hewitt, P. Physics Teacher *27 (1989): 112.*

78. Sailing in Calm Air!

Raise the sail, and the wind created by the ship's movement with the current will help push the boat forward. You cannot sail directly into the wind, but must sail a zigzag course against the headwind, just as sailboats normally do.

Bradley, R. C. "Problem: Sailing Down the River." American Journal of Physics *64 (1996): 686, 826.*

79. The Impossible Dream

Yes, under very specific conditions. The air molecules driven forward by the fan must reach the sail and bounce off with some backward component to their velocity. The standard argument invokes the conservation of momentum. Obviously, the boat with the fan plus the air between the fan and the sail do not constitute a closed system.

A single air molecule at rest in front of the fan (or a fixed volume of air with average velocity zero) can be struck by the fan blade and driven forward with vector momentum p. The fan blade—and the boat connected to the fan—gain the same momentum $-p$ in the opposite direction. In the upper limit of momentum transfer, the molecule will make an elastic collision with the sail and bounce backward with momentum $-p$. The molecule's change in momentum is therefore $-2p$, so the sail—and the boat—receive momentum $+2p$ in the forward direction. Adding together the two momentum changes imparted to the boat, its total momentum change is $-p + 2p = +p$, a net increase in the forward direction.

Note that in the limit of the molecule sticking to the sail, the net result is zero momentum for the boat. The second citation below gives an account of the conditions required for a boat on an airtrack and demonstrates the successful operation.

Clark, R. B. "The Answer is Obvious, Isn't It?" Physics Teacher *24 (1986): 38–39.*

Hewitt, P. "Figuring Physics." Physics Teacher *26 (1988): 57–58.*

Martinez, K., and M. Schulkins. "Letters." Physics Teacher *24 (1986): 191.*

80. Lifting Power of a Helium Balloon

No. The helium-filled balloon does much better than expected. From Newton's second law, you calculate the net force in the vertical direction, the buoyant force upward minus the weight downward. The buoyant force upward equals the weight of the displaced volume of air, while the total weight is the weight of the gas inside the balloon plus the weight of the balloon skin plus the weight of the payload.

The lifting ability is the buoyant force of the air minus the weight of the gas in the balloon, a quantity proportional to their difference in molecular weight. The average molecular weight of air at sea level is 28.97, producing a difference of 24.97 for helium compared to a difference of 26.97 for hydrogen. The relative lifting ability of helium is the ratio of $24.97/26.97 = 0.926$; that is, helium is 92.6 percent as good for lifting.

Burgstahler, A. W., T. Wandless, and C. E. Bricker. "The Relative Lifting Power of Hydrogen and Helium." Physics Teacher *25 (1987): 434.*

Lally, V. E. "Balloons for Science." Physics Teacher *20 (1982): 438.*

81. Reverse Cartesian Diver

The bottle cannot have a circular cross section. Squeezing the noncircular-cross-section bottle across the wider direction reduces the water pressure, and the diver is pushed upward. Squeezing a circular-cross-section bottle would increase the pressure and keep the diver on the bottom.

Brandon, A. "A Beautiful Cartesian Diver." Physics Teacher *20 (1982): 482.*

Butler, W. A. "Reverse Cartesian Diver 'Trick.'" American Journal of Physics *49 (1981): 92.*

Wild, R. L. "Ultimate Cartesian Diver Set." American Journal of Physics *49 (1981): 1185.*

82. Cork in a Falling Bucket

The cork will still be at the bottom because the bucket, the water, and the cork all fall with exactly the same acceleration g (neglecting the air resistance effects). One might expect the buoyant force of the water to push the cork upward to the surface, but in free fall the buoyant force is zero.

83. Immiscible Liquids

After the liquids separate, the central column weight is less; therefore the pressure at the bottom is less. The sloping walls of the bottle also push downward less to complete the argument.

Arons, A. B. Teaching Introductory Physics. *New York: John Wiley & Sons, 1997, pp. 327-328.*

84. The Hydrometric Balance

Surprisingly, the tube maintains its equilibrium position in the liquid, and any vertical oscillations of the platform have no effect on the position! When the platform is accelerating upward, the extra buoyant force of the liquid just balances the extra downward force resulting from the acceleration. Likewise for any downward acceleration.

Weltin, H. "Mechanical Paradox." American Journal of Physics *34 (1966): 172.*

85. Child with a Balloon in a Car

The air inside the car will tend to continue its straight-line motion momentarily, so the air pressure inside the car will be slightly higher on the outside radius of the turn. The balloon will then be pushed to the right, toward the inside of the turn.

Lehman, A. L. "An Illustration of Buoyancy in the Horizontal Plane." American Journal of Physics *56 (1988): 1046.*

86. The Reservoir behind the Dam

No. The depth of the water immediately behind the concrete dam is all that matters, because the water pressure depends upon the *depth* of the water h and its density ρ. The total pressure P at depth h in the water is $P = P_0 + \rho gh$, where P_0 is the atmospheric pressure. The total amount of water in the reservoir behind the dam is not relevant. Nor is the amount of water in the river above the dam. A thin 10-meter-high film of water contacting the dam requires the same dam strength as a wide 10-meter-high lake.

87. Finger in the Water

Yes. The pan with the bucket will go down. The water exerts a buoyant force on your finger and, by Newton's third law, the finger exerts an equal and oppositely directed force on the water that is transmitted to the bottom of the bucket, to the pan, and to the balance, causing it to dip.

88. The Passenger Rock

The water level remains unchanged. The same volume of water is displaced in both orientations.

Hewitt, P. "Figuring Physics." Physics Teacher *25 (1987): 244.*

89. Archimedes in a Descending Elevator

No. First assume that we can ignore surface tension effects. Then note that both vertical forces—the upward buoyant force and the downward weight of the block—are directly proportional to the gravitational force. Decreasing the vertical acceleration via $g - a$ decreases the weight and the buoyant force equally, so the block maintains its position in the water.

90. Three-Hole Can

The solution shown in the figure is incorrect. The water stream from the middle hole would go the farthest horizontal distance, and the other two streams would go the same horizontal distance.

The horizontal distance traveled by a stream is given by $s = vt$, where v is the horizontal exit velocity from the hole and t is the time of flight, which is the same time interval as the free-fall time (ignoring air effects). Let H be the constant height of the water column, with the holes at heights $H/4$, $H/2$, $3H/4$. One can derive Torricelli's law from the law of conservation of energy: The kinetic energy $\frac{1}{2} mv^2$ of the efflux stream from the hole equals the difference in potential energy mgh, where h is the distance below the water head. Thus $v = \sqrt{2gh}$. The time of free fall t from the height $(H - h)$ is simply $t = \sqrt{2(H - h)/g}$. Multiplying, one obtains the expression $s = 2\sqrt{h(H - h)}$, which has a maximum at $h = H - h$, or $h = H/2$. Substitutions will verify that the other two streams should hit together on the table surface.

91. The Laundry Line Revealed!

The most obvious explanation—that gravity draws the water down and out of the fabric—is wrong. The water in the fabric is held in the spaces between the threads by electric forces (i.e., capillary action), and the gravitational force cannot dislodge this water. Gravity is involved in the real explanation, but only in a secondary role.

The slow evaporation of water into the air next to the garment cools this contact air, which is now more dense than the surrounding warmer air. This more dense air moves downward across the face of the cloth, and the moving air soaks up the evaporated water molecules, becoming more saturated as it sinks. The uptake of water vapor will be greatest at the top and less farther down because the more saturated the air becomes, the less its ability to soak up water molecules. So the garment dries from the top down.

Hansen, E. B. "On Drying of Laundry." SIAM Journal on Applied Mathematics *52 (1992): 1360.*

"Mathematics of Laundry Unveiled." Science News *142 (1992): 286.*

92. Pressure Lower than for a Vacuum!

For liquids, attractive forces between molecules can make the pressure negative. One usually thinks about pressure in gases, which can only have positive pressures resulting from repulsive forces related to collisions. But liquids can have negative, zero, or positive pressures. By the way, water at 0 Pa has molecular kinetic energy, whereas the vacuum has none.

Kell, G. S. "Early Observations of Negative Pressures in Liquids." American Journal of Physics *51 (1983): 1038.*

Kuethe, D. O. "Confusion about Pressure." Physics Teacher *29 (1991): 20–22.*

93. Canoe in a Stream

Probably not. As the canoe approaches the stream narrowing, the water flows faster at the front end than at the back end of the canoe. The result will be a canoe oriented parallel to the flow of the water. A small angle deviation from the flow direction will encounter a restoring torque at the front greater than the opposite torque at the rear.

Crane, H. R. "Stretch Orientation: A Process of a Hundred Uses." Physics Teacher *23 (1985): 304.*

94. Water Flow Dilemma

The water flows from the left into the right graduated cylinder, and eventually the water levels will match. The system responds to the difference in pressures. Many people try to utilize the difference in weights of the water columns to predict the correct behavior. If their argument were true, the flow would be in the other direction.

95. Iron vs. Plastic

As the air is removed quietly so that no convection currents arise, both spheres are buoyed up less, but the decrease is greater for the larger, plastic ball, so the plastic ball moves downward.

96. Iron in Water

The submerged sphere is subject to a buoyant force equal to the weight of the water in the volume displaced by the sphere. Let us call this weight of water *w*. One might be tempted to say that to restore equilibrium, a weight *w* should be added to the pan with the stand. However, according to Newton's third law, the force with which the water in the container acts on the submerged sphere is exactly equal to the force with which the sphere acts on the water in the opposite direction. Hence, as the weight of the pan with the stand decreases, the weight of the pan with the container increases. Therefore, to restore balance, a weight equal to 2*w* must be placed on the pan with the stand. By the way, the tip of the balance does not indicate unequal torques. Two objects can balance at any angle of tip.

97. Paradox of the Floating Hourglass

The paradox arises because the buoyant force should be the same at all times when the hourglass is totally submerged, but the behavior seems to contradict this statement.

When the unit is turned over and the hourglass is inverted at the bottom, its slight tipping angle pushes its glass against the glass cylinder where the contact friction and the surface tension of the water prevent upward movement. When enough sand has fallen to the bottom of the hourglass, the torque that tips the hourglass is reduced significantly. Then the upward buoyant force becomes greater than the opposing forces—weight, contact friction, and surface tension—so up it goes.

Gardner, M. Scientific American 215 (1966): 96.

Reid, W. P. "Weight of an Hourglass." American Journal of Physics *35 (1967): 351.*

98. Open-Ended Toy Balloon

First turn the balloon inside out. Heat about 5 milliliters of tap water in a 500-milliliter Florence flask until the water boils rapidly. Most of the air inside the flask will be replaced by hot air with water vapor. Put on some rubber gloves for safety, and then quickly stretch the mouth of the balloon over the mouth of the flask. This latter procedure must be done very quickly to prevent much outside air from reentering the flask.

As the water stops boiling, the flask cools and the water vapor pressure will drop rapidly. The outside air will inflate the balloon inside the flask.

Louvière, J. P. "The Inscrutable, Open-Ended Toy Balloon." Physics Teacher *27 (1989): 95.*

99. Response of a Cartesian Diver

The sharp blow to the countertop sends a compressional shock front into the bottom of the container through the water to the diver to momentarily reduce its air volume. If the buoyancy was originally marginal, the diver will coast downward to the bottom.

Orwig, L. P. "Cartesian Diver 'Tricks.'" American Journal of Physics *48 (1980): 320.*

100. Perpetual Motion

Each liquid exerts forces only perpendicular to the surface of the cylinder, so no torques are present. There is no rotation.

Miller, J. S. "An Extraordinary Device." Physics Teacher *17 (1979): 383.*

101. Double Bubble

The larger soap bubble will get larger and the smaller bubble will get smaller because the air pressure inside a soap bubble decreases with increasing radius. Roughly, for a spherical bubble of radius R, the surface tension force $2 \pi RT$ is equated to the force provided by the inside air pressure $4 \pi R^2P$, leading to the pressure $P \propto 1/R$. In the case of two balloons, the larger one will force the air into the smaller until their sizes become equal.

102. The Drinking Straw

Nothing happens! The water remains inside the straw. The pressure inside the straw is less than atmospheric. (You must ensure that the hole is large enough so that surface tension plays a minor role only.)

103. Hot-Air Balloon

The real explanation depends upon the density of the balloon with respect to the density of the surrounding air. The air inside the balloon adds weight to the balloon, whether hot or cool. What the hotter air does is push out the walls of the balloon more to increase the volume and thereby decrease the density. Then up it goes!

104. Improving the Roman Aqueduct

Yes, the water can flow over a hill that is higher than the water source head. This kind of device is called a siphon. A siphon works best if the flow can remain laminar—that is, with no turbulence. This condition can be met by making the cross section of the tube vary with elevation. Energy considerations dictate that the water flows slower at the higher elevations along the journey, so higher elevations must have a bigger cross section to maintain the same flow rate.

Benenson, R. E. "The Hyphenated Siphon." Physics Teacher *29 (1991): 188.*

Ansaldo, E. J. "On Bernoulli, Torricelli, and the Siphon." Physics Teacher *20 (1982): 243.*

105. Barroom Challenge

Use one of the additional stirrers to blow air into glass A at any point where glasses A and B meet. Some water will trickle out of glass A into glass C as the air occupies the uppermost volume in the glass.

Schreiber, J. T. "Barroom Physics." Physics Teacher *13 (1975): 361, 378.*

106. Tire Pressure

The tire pressure will be nearly the same in both cases. Even though the tire volumes are different in the two cases, this difference is small. The air pressure is slightly more when the tire helps support the weight of the car. The stiff tire sidewall actually provides much of the support for the car.

*107. The Siphon

For this analysis of the siphon, we consider the ideal case: a nonviscous and incompressible liquid, with no dissipation of energy, and a large liquid container with a cross-sectional area very large with respect to the siphon tube diameter so that the liquid level is essentially constant. The best approach is to realize that siphon operation depends upon a dynamical model, not a static one. However, one can use the static model to explain how to get the siphon started.

Getting it started: The siphon tube has end *A* in the liquid and end *F* outside. If end *F* is slightly lower than end *A*, and one has drawn fluid into the tube to completely fill it, then the pressure inside the end *F* is slightly greater than atmospheric pressure, so fluid flows through the siphon. This flow continues until the pressure inside end *F* reaches the atmospheric pressure, the pressure decrease occurring because the level inside the liquid container is decreasing its head. Then, if one does not lower end *F*, the flow will eventually cease. In our ideal case of a very large liquid container, the flow continues forever.

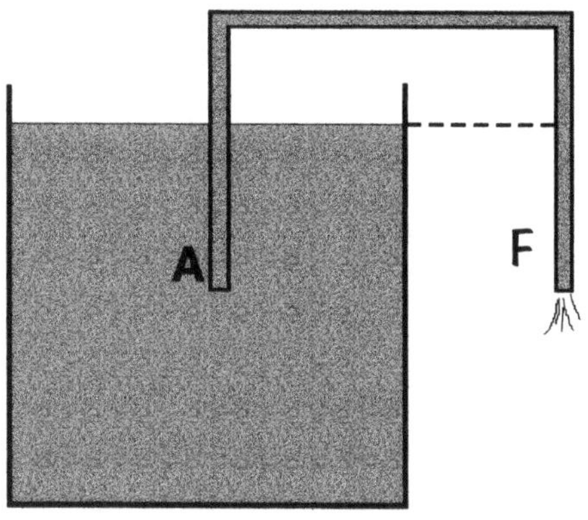

The dynamic theory for the ideal case and *no turbulence:* Two equations must be combined to completely understand the siphon operation. First, one has the Bernoulli equation for points along a streamline, which accounts for energy conservation:

$$p_a + \rho g h_a + \tfrac{1}{2}\rho v_a^2 = p_b + \rho g h_b + \tfrac{1}{2}\rho v_b^2 ,$$

where a is one point on the streamline and b is another, with p the pressure, h the height, v the fluid velocity, ρ the fluid density, and g the acceleration of gravity. Second, there is the equation of continuity for the volume of fluid per second $Q/t = Av$, where A is the uniform cross-sectional area of the siphon tube. In the real case with viscosity, a third equation—Poiseuille's equation—would be needed.

One now applies the above equations to the siphon. At all points inside the tube the pressure is less than the surrounding atmospheric pressure p_0, except perhaps at end *F.* For example, consider point a to be inside the tube at the same height as the liquid surface for the large container, and let point b be inside the tube at the top of the siphon at distance h above this liquid surface. Then one obtains $p_0 = p_b + \rho g h_b + \tfrac{1}{2}\rho v_b^2$, or $p_b = p_0 - \rho g h_b - \tfrac{1}{2}\rho v_b^2$, showing the internal pressure less than the outside air pressure. Note that p_0 can be set to zero later if we want to operate in a vacuum.

One now shows that a pressure difference just outside end *A* in the liquid to just inside end *A* in the tube is the "engine" that drives the siphon. With a tube of uniform bore, the velocity of flow v will be the same throughout the tube. Outside end *A* the pressure is $p_0 + \rho g h_a$, while the pressure just inside end A is $p_0 + \rho g h_a - \tfrac{1}{2}\rho v_a^2$, where the minus sign is correct. Thus the pressure drops by $\tfrac{1}{2}\rho v_a^2$ across the tube entrance.

Notice that the motion of the fluid is critical in explaining the operation, so static models are incomplete. Also, the atmospheric pressure p_0 cancels out and therefore does not drive the liquid up the tube.

Ansaldo, E. J. "On Bernoulli, Torricelli, and the Siphon." Physics Teacher *20 (1982): 243.*

Benenson, R. E. "The Hyphenated Siphon." Physics Teacher *29 (1991): 188.*

*108. Reverse Sprinkler

Conservation of angular momentum dictates that the sprinkler should rotate oppositely in the two opposite modes, and it does. One cannot use time reversal analysis here because in the forward mode, the normal sprinkler operation, the water pressure is lower in the surrounding medium at the ends of the nozzles. Running a film backward of this normal operating mode does not match the reverse mode of operation because the pressure regions inside the nozzles do not get reversed. The only way water can enter the sprinkler at the nozzle ends is by having a higher pressure in the surrounding medium than inside. There is a further complication: In the inverse mode the entering water comes from all directions, so only water actually in the nozzles contributes to the angular momentum of the system, whereas in the normal forward mode all the water exiting contributes.

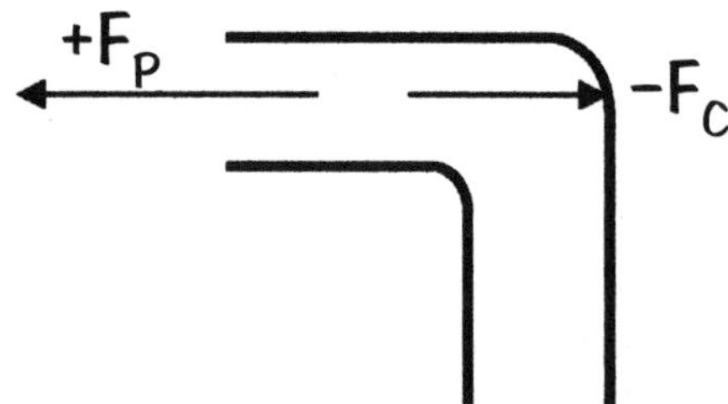

For simplicity, consider the water to have no viscous drag inside the nozzle. First consider the azimuthal (rotational forces not along the radial direction) forces acting on the entering water: $-F_p$, a liquid pressure difference clockwise and inward at the nozzle end times the area of the nozzle orifice, and the force $+F_c$, which changes the water flow from azimuthal to radial at the bend inside. These two forces point in opposite directions. Second, consider the corresponding reaction forces acting on the nozzle in the azimuthal direction: $+F_p$ pointing counterclockwise and outward at the orifice and $-F_c$. Before reaching a steady-state flow condition, $+F_p$ is greater than $-F_c$, so the inverse sprinkler rotates oppositely in water to the forward sprinkler.

The inverse sprinkler operating in air behaves differently (!), both cases being compared in the article by Collier, Berg, and Ferrell.

Berg, R. E., and M. R. Collier. "The Feynman Inverse Sprinkler Problem: A Demonstration and Quantitative Analysis." American Journal of Physics *57 (1989): 654–657.*

Collier, M. R.; R. E. Berg; and R. A. Ferrell. "The Feynman Inverse Sprinkler Problem: A Detailed Kinematic Study." American Journal of Physics *59 (1991): 349–355.*

Schultz, A. K. "Comment on the Inverse Sprinkler Problem." American Journal of Physics *55 (1987): 488.*

*109. Spouting Water Droplets

Energy couples into the sliding cup as it is driven by the hand as a relaxation oscillator—that is, in a slip-stick mode of vibration as the nearly flat bottom of the cup catches and releases on the finished wooden surface. With the good coupling at the right sliding speed, standing waves are created almost immediately on the surface of the liquid. A continued steady push of the cup provides enough upward movement at wave crests in two dimensions to cause drops of water to break from the liquid surface and be projected high above the cup.

Keeports, D. "Standing Waves in a Styrofoam Cup." Physics Teacher *26 (1988): 456–457.*

Chapter 4 Fly like an Eagle

110. Vertical Round Trip

Sometimes the journey will take longer coming down than going up. For example, a paper glider thrown vertically upward can glide downward ever so slowly. But for many objects, the total travel time will be less. A ball thrown upward, or a dart thrown upward, does not go as high with the same initial velocity, and the round-trip time is less than for free fall. One also can show that a ball takes more time to fall than to go up, because at all heights the downward speed will be less than the upward speed at the same height.

Pomeranz, K. B. "The Time of Ascent and Descent of a Vertically Thrown Object in the Atmosphere." Physics Teacher *7 (1969): 507–508.*

111. It's a Long Way to the Ground!

No, the terminal velocity does not depend upon the altitude of the drop. Although objects dropped from high altitudes—3 kilometers or more—can reach velocities as much as several hundred meters per second, their terminal speeds are all the same near the ground, where the air resistance force is proportional to the square of the speed. Can a second identical sphere pass the first sphere? Only if the air effects retarding its motion are decreased by the first sphere as it falls so that the second sphere can fall faster.

Shea, N. M. "Terminal Speed and Atmospheric Density." Physics Teacher *31 (1993): 176.*

112. Galileo's Challenge Revised

The buoyant forces are equal, but the bowling ball weighs much more. Applying Newton's second law, the bowling ball has the greater initial downward acceleration, and this condition is maintained all the way down. The bowling ball experiences a slightly greater air resistance force on the way down because it is moving faster, but the plastic ball never catches up.

Nelson, J. "About Terminal Velocity." Physics Teacher *22 (1984): 256–257.*

Toepker, T. P. "Galileo Revisited." Physics Teacher *5 (1967): 76, 88.*

Weinstock, R. "The Heavier They Are, the Faster They Fall: An Elementary Rigorous Proof." Physics Teacher *31 (1993): 56–57.*

113. Falling Objects Paradox

The dropped object hits first! For the object fired horizontally, Newton's second law produces an acceleration in the vertical direction $a = -g + BV v/m$, where g is the acceleration of gravity, B is a constant for the air viscosity, V is the large instantaneous velocity magnitude of the object, v is its component value in the vertical direction, and m is the mass. The second term tells us that the magnitude of the vertical acceleration of the fired object is less than for the dropped object, whose vertical acceleration is $-g + B v v/m$ because $V >> v$.

When considering the effect of the curvature of the earth on the time of flight for the horizontally shot cannon ball, one can use several different approaches. A simple one examines the two limiting cases: (1) zero initial horizontal velocity, so the cannon ball drops just like the other one; and (2) the cannon ball exits with a horizontal velocity that produces a circular orbit (or nearly so) to make the time of flight extremely large. All other cases of interest for a collision with the earth lie between these two. Therefore, the dropped cannon ball hits first. One could also analyze the motion by considering the effect of centrifugal force on the radial fall of the object.

114. Iceboat

Yes. Even though the iceboat is constrained to move in the direction of its runners, this behavior gives it a stability to the sideways push of the wind. Like the normal sailboat in water, the iceboat can move much faster than the wind speed driving it. One simply positions the sail properly when the boat is tacking into the wind so that there is a small forward component of the wind force on the sail in addition to its sideward force component. Some iceboats can achieve speeds of up to two to three times faster than the wind speed.

115. The Flettner Rotor Ship

The rotation direction of the vertical cylinder is important. To utilize the Bernoulli effect, you must create a lower pressure in front of the rotating cylinder than the pressure behind the cylinder. In the given conditions, with the ship headed west and the wind from the south, the cylinder should be rotated clockwise from above. To the front, the velocity of the air is added to the rotational tangential velocity of the cylinder. To the back, the velocity of the air and the rotational tangential velocity are opposite and subtract. The Bernoulli effect dictates a lower pressure at the front, and the ship is pushed forward by the effects of the wind.

Barnes, G. "A Flettner Rotor Ship Demonstration." American Journal of Physics *55: 1040–1041.*

116. The Lift Force Is Greater, Isn't It?

In a stable, constant rate of climb the lift force is *less than* the weight of the airplane. The thrust has an upward component, which adds to the lift, to balance the weight.

Most people expect that airplanes climb because the lift exceeds the weight—an incorrect intuitive approach. The plane is not accelerating in any direction in the simplest case. Along the line perpendicular to the wing (i.e., along the lift direction), the forces sum to $L - W \cos \alpha = 0$, where L is the lift, W is the weight, and α is the climb angle. Therefore, for any climb angle, the lift force is less than the weight.

Flynn, G. J. "The Physics of Aircraft Flight." Physics Teacher *25 (1987): 368–369.*

117. Floating Rafts

Throughout the flowing water, viscous forces are acting to speed up or slow down layers of water in the river. The water near the banks and near the river bottom experience the resistive drag effects of the nearly still water in contact with these solid surfaces. At the same time, the water flowing farther from these surfaces is trying to accelerate the nearly still waters via viscous effects. A boundary layer forms—that is, there develops a layer of retarded flow. Eventually a steady-state flow rate profile usually develops, with the velocity increasing inward toward the center of the river and upward from the bottom, reaching a maximum speed near the center just below the water surface.

The maximum flow rate occurs a short distance below the surface, because the air above the water surface actually provides a drag force on the water. Therefore, a more heavily loaded raft, being deeper in the water, will be pushed by a faster current and will float faster than a lightly loaded raft.

118. Dubuat's Paradox

The water resistance is usually less when the stick is held in a moving stream. In a liquid there will be friction drag and form drag. The friction drag for blunt objects is insignificant when compared to the amount of form drag. The water molecules in contact with the stick in the stream will be nearly stationary, and there is a boundary layer of retarded flow from the stick to a significant distance away, several stick diameters.

Flowing streams are somewhat turbulent, and this turbulence induces a transition in the boundary layer surrounding the stick. As a result, the slow-moving boundary layer receives extra kinetic energy from the free stream and follows farther around the stick without separation than normally occurs. The form drag is reduced and so is the total drag, since the friction drag is insignificant here.

119. Airfoil Shapes in the Airstream

At a slow air speed of 300 kilometers per hour, orientation (a), with the rounded edge forward, offers less air resistance. For this "high-speed flight" through a low-density fluid, the Reynolds number is $R >> 1$, so the viscous forces have only a minor influence.

120. Airfoil Shapes in the Waterstream

Viscous forces dominate for reasonable water speeds such as 20 knots. The Reynolds number $R < 1$ and orientation (b) offers less water resistance.

121. Wire vs. Airfoil

The airfoil. Its streamlined shape, although ten times thicker, produces slightly less drag than the round wire because it prevents turbulence on the backside of the shape when air flows past. A turbulent region behind would be a lower-pressure region, producing a net force backward on the object, effectively contributing to the flow resistance. The airfoil shape reduces the formation of turbulence significantly, compared to the round cross-section wire.

122. Hole-y Wings!

The air flow over and under a typical wing breaks up into turbulence, and the drag increases. By making the holes and "sucking" in the turbulent air with a pump through those holes, the drag is reduced significantly. Less drag means less fuel required and a less expensive operational cost.

Browne, M. W. "New Plane Wing Design Greatly Cuts Drag to Save Fuel." New York Times, *September 11, 1990, pp. C1, C9.*

123. Frisbee Frolics

When the center of lift is ahead of the center of gravity, a slight tilt that moves the front of the Frisbee (and the center of lift) upward or downward would become an unstable condition if the object were not spinning. With the angular momentum of the spin, this tilt produces a slow precession of the spin axis, analogous to the behavior of a gyroscope. The resulting wobble creates a lot of turbulence and increases the flight drag, shortening the flight distance.

Crane, R. "Beyond the Frisbee." Physics Teacher *24 (1986): 502–503.*

124. Aerobie Frolics

The Aerobie solves some of the aerodynamic problems of the Frisbee noted above. The Aerobie has an outer edge—a rim—that acts as a "spoiler" to make the airflow break away from the surfaces of the *leading* part of the wing, introducing some turbulence. This leading part loses some lift, but now the center of lift is very close to the center of gravity instead of being ahead of it. Overall, less turbulence occurs than before, when there was more wobble associated with precession. Hence, less drag means the Aerobie can go a farther distance than the Frisbee. Now, if one could eliminate the turbulence altogether!

Crane, R. "Beyond the Frisbee." Physics Teacher *24 (1986): 502–503.*

125. Kites I

The angle between the kite face and the wind direction must be adjusted for best performance. This angle of attack should be smaller for greater wind speeds; otherwise the kite becomes unstable and may even break up. As the kite moves up to greater heights, the wind speed usually increases and the angle of attack is no longer near optimum. The spring or rubber band stretches in response to the force of the wind to adjust the angle of attack. The kite's tail can help reduce stability problems, but there is a limit to its effectiveness.

Walker, J. "Introducing the Musha, the Double Lozenge, and a Number of Other Kites to Build and Fly." Scientific American *257 (1978): 156–161.*

126. Kites II

The drogues are tapered, so air entering at the upwind side speeds up and exits on the downwind side. This outgoing stream, moving faster than the surrounding airstream, helps maintain the proper orientation of the drogues so that they help dampen any lateral excursions. In other words, any misalignment of the drogues moves the exit end into a higher-pressure region, which pushes it back.

Walker, J. "Introducing the Musha, the Double Lozenge, and a Number of Other Kites to Build and Fly." Scientific American *257 (1978): 156–161.*

127. Parachutes

An unvented parachute alternately creates vortexes on opposite sides of the parachute, and the parachute responds by swinging more and more. As the air passes the edges, the pressure in the vortex is lower than the ambient air pressure, the swinging begins, and the swinging amplitude increases with each periodic impulse. The vents break up the vortexes to reduce the swinging.

128. Strange Behavior of a Mixture

This mixture is an electrorheological liquid, one whose viscosity is affected by electric fields. Neither the oil nor the cornstarch is electrically conducting, but they are dielectrics. For the liquid to pour, the oil must flow and the particles of cornstarch must move with the oil and past one another.

The electric field polarizes the cornstarch particles, and strings of cornstarch in the oil form to restrict the movement of the oil. These strings cannot move around each other smoothly, so the liquid becomes more viscous. You can also move the charged object up close to the surface of the liquid as it sits in a glass to see the formation of a temporary dent in the surface.

Haase, D. "Electrorheological Liquids." Physics Teacher *31 (1993): 218–219.*

129. Catsup

Catsup is a thixotropic liquid, one whose viscosity reduces with flow speed. Apparently the flow causes long chains and strands to align with the flow direction to reduce the flow resistance.

130. Coiled Garden Hose

When the water is poured in via the funnel, filling the first loop, some will fall to the bottom of the second loop. An air trap will form at the top of the first loop. If more water is poured into the funnel, a few more air traps may form at the tops of the loops until the pressure of the water column beneath the funnel is insufficient to push water upward in the loops to eliminate the air traps. Past that point of operation, no more water will enter the hose. And none will exit the other end.

131. Flow from a Tube

A non-Newtonian fluid, such as a long-chain polymer fluid, spreads outward upon first exiting the orifice. The long-chain molecules get tangled together, and their entanglement increases and occupies more volume, in contrast to normal fluids.

Bird, R. B., and C. F. Curtiss. "Fascinating Polymeric Liquids." Physics Today *37 (1984): 36–43.*

Walker, J. ". . . More about Funny Fluids." Scientific American *259 (1980): 158–170.*

132. Spheres in a Viscous Newtonian Liquid

The second sphere catches up to and collides with the first. Each sphere slows in the same way. If they were very small, a separation distance would always be maintained. But when they have a physical size, they will eventually touch if the fluid path is long enough.

Bird, R. B., and C. F. Curtiss. "Fascinating Polymeric Liquids." Physics Today *37 (1984): 36–43.*

133. Spheres in a Viscous Non-Newtonian Liquid

There are two solutions for spheres moving through a non-Newtonian liquid. If the second sphere is dropped very soon after the first one, the second will approach the first and collide, for the same reason as in the Newtonian liquid when the spheres have a physical size.

However, if the delay in the release of the second sphere is longer than a critical time interval, the spheres will move apart while falling. The motion of the first sphere through the fluid has increased the viscosity of the liquid for the second sphere. The shearing done by the first sphere increased the viscosity.

Bird, R. B., and C. F. Curtiss. "Fascinating Polymeric Liquids." Physics Today *37 (1984): 36–43.*

134. Animalcules in $R < 10^{-4}$

The Reynolds number of a fluid is measured empirically and is important for determining when laminar flow could become turbulent. At such very low Reynolds number values, the flow is laminar, and every action reverses quite well. The direction of time is practically meaningless as far as motion is concerned. The elapsed time on the forward stroke and a different time on the reverse stroke make no difference. Only configuration changes are effective, so if the animal tries to swim via reciprocal motion, it cannot go anywhere. Fast or slow, the animal retraces its trajectory back to its starting location.

The animalcules that do swim have a flagellum that turns like a corkscrew or have a flexible deformable oar mechanism, neither of which practices reciprocal motion.

Purcell, E. M. "Life at Low Reynolds Number." American Journal of Physics *45 (1977): 3–11.*

*135. Lift without Bernoulli

The lift created by the wing can be explained by applying Newton's second law in accounting for the deflection of the airflow upward and downward by the whole wing. We are concerned here with the changes in momentum for the airflow deflection downward versus the changes in momentum for the airflow deflected upward during each second. Recall that the momentum is the product of the mass and the velocity, and that for the wing we primarily have a change in the velocity. When the downward change in momentum each second exceeds the upward, there is lift.

The amount of lift force depends upon the speed and density of the air, the shape of the wing, and the angle of attack. Most airplane wings could be turned upside down and still produce lift over a wide range of conditions. Besides, the airflow over and under a wing surface is quite complicated, with turbulence and other effects being part of the behavior—certainly not conditions conducive to Bernoulli flow, which assumes laminar flow. For a more satisfactory explanation, all these effects can be lumped into one package by considering just Newton's second law and the changes in the momentum of the airflow.

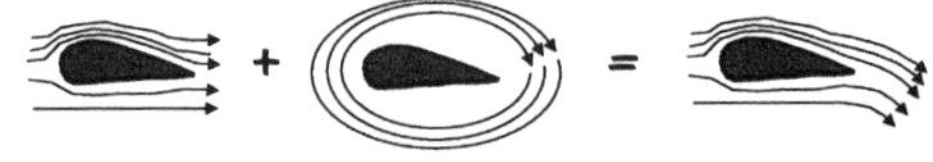

There is another approach toward understanding the aerodynamic lifting force in terms of the Kutta-Joukowski equation, which relates the lifting force and the flow of downward momentum produced by the wing acting as an airfoil, using the concept of circulation around the wing. For a circulation Γ, the lifting force $F = \rho v \Gamma$, where ρ is the fluid density and v is the streamline flow velocity. One can show that the circulation is a constant for all closed curves around the airfoil, then one can calculate the velocity differences between the upper side and the bottom of the wing profile, and finally one can apply Bernoulli's law to determine the pressure differences. These pressure differences then drive the air over and under the wings, not the other way around, as is often presented in textbooks!

Definitely, the concept of circulation helps in the understanding of the Kutta-Joukowski formula as applied to airfoils by simply calculating the flow of downward momentum induced by the airfoil. However, the circulation is not the physical reason for the lift force, and one does not need to consider circulation to explain the origin of lift.

In summary: Lift occurs if and only if the wing, by its profile and by its angle of attack, gives the airflow a net downward momentum.

Weltner, K. "Bernoulli's Principle and Aerodynamic Lifting Force." Physics Teacher *28 (1990): 84–86.*

———. *"A Comparison of Explanations of the Aerodynamic Lifting Force."* American Journal of Physics *55 (1987): 50–54.*

*136. Storm in a Teacup

This interesting behavior of tea leaves has intrigued many people, including Albert Einstein, who published a paper on this phenomenon in 1926 that you can read in A. P. French, *Einstein: A Centenary Volume* (Cambridge, Mass.: Harvard University Press, 1980).

Without friction along the cup walls and bottom, any minuscule rotating fluid volume has no outward radial motion. The pressure in the liquid increases outward from the central axis as the liquid volume is accelerated inward to maintain its rotation radius.

But there is friction between the bottom layer of fluid and the bottom of the cup. This friction reduces the rotation speed and the pressure difference between the fluid near the wall and the fluid in the center. This reduction is much less higher up in the liquid. As a result, liquid is pushed down along the wall, then radially inward to the center of the cup, then upward at the center, and then outward near the top.

The tea leaves are carried along to the center, but the total upward force of the fluid flow plus the buoyant force are not great enough to carry them upward against their weight.

Davies, P. "Einstein's Cuppa." New Scientist *154 (1992): 52*

Smith, J. "Twirling Tea Leaves." New Scientist *154 (1992): 53.*

Walker, J. "Wonders of Physics That Can Be Found in a Cup of Coffee or Tea." Scientific American *256 (1977): 152–160.*

*137. Smoke Rings I

The movement of the smoke ring involves a connection between force and velocity, not force and acceleration as given in Newton's second law.

This force-velocity relationship is derivable by applying Newton's second law, but the details are complicated. Let's take a simple view of the behavior. For the opposite sides of the smoke ring, the particles are rotating in opposite directions, as shown. Although seemingly separate, these opposite sides influence each other. Specifically, the rotation of the smoke in the top vortex causes the smoke in the bottom vortex to move to the right. In exactly the same manner, the bottom vortex causes the smoke in the top vortex to move to the right. This argument is true for any opposite sections of the toroidal smoke ring.

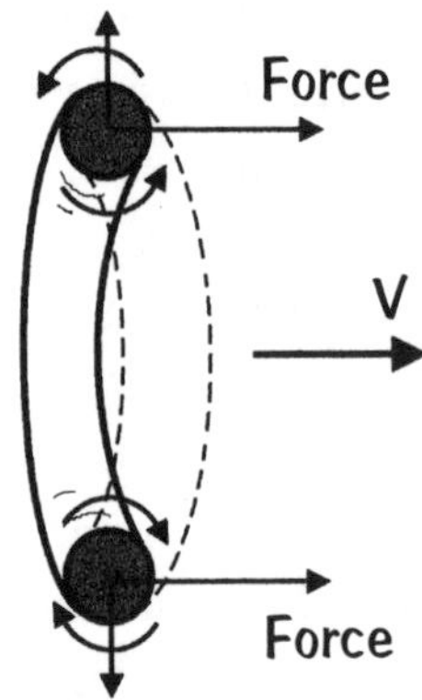

When would a smoke ring expand and when would it shrink? The diagram shows that if a force is applied at right angles to the plane of the ring, the toroidal axes of the two opposite vortices are pushed into regions where the fluid rotates so that the vortex axes are driven outward to increase the ring diameter. Simultaneously, the forward movement of the smoke ring decreases. Why? If the applied force is in the other direction, the ring shrinks, and its forward motion speeds up. Under ideal conditions, two nearby smoke rings can act upon each other to pass through one another!

*138. Smoke Rings II

The two coaxial smoke rings moving in the same direction actually attract each other, analogous to two electrical current loops of the same current directions. The vortices around one smoke ring act on the vortices around the other ring to sweep them closer to each other. As a result, the trailing ring is accelerated, and the leading ring is decelerated. All work better when the trailing smoke ring initially has a much greater speed than the leading one. However, multiple passages of smoke rings are difficult under the best of circumstances.

Chapter 5
Good Vibrations

139. Conch Shell

The cavity inside the conch shell acts as a resonator for any sounds entering from the ambient air or from the human ear or by contact transmission through the head bones and skin to the shell material. One can experience this same phenomenon of cavity resonance by snapping one's middle finger with the thumb when the hand is closed and when the hand is open. The closed-hand finger-snapping sound is significantly louder. Likewise, sounds from the blood rushing by the ear as well as the ambient ocean sounds produce quite an interesting effect when heard emanating from the conch shell.

140. Hearing Oneself

The difference is real. Your voice sounds thinner and less powerful to others than to you because you hear your own voice through skull bone conduction as well as via air conduction. You can verify the difference: hum with closed lips, then stopper your ears with your fingers, and the hum will be louder! In air-conducted sounds, most of the vibrational energy goes into frequencies above 300 hertz, with only very little going into the lower-frequency sounds.

141. A Rumble in the Ears

This rumbling sound at about 23 hertz originates in the muscles in your arms and hands. The actin and myosin microfilaments in the muscles are continually stretching slightly and relaxing slightly. Each small movement involves some rubbing of one muscle over another to produce sounds that are transmitted along the forearm bones to the hand. You can verify their source by first listening with your arms somewhat relaxed to establish a baseline sound intensity, then tensing your fist and forearms to hear the sound intensity increase manyfold.

If a gorilla did the same experiment, listening to her muscles, the rumbling sound intensity may be quite a bit louder, because a gorilla's finger muscles in the hand itself are quite substantial, in contrast to the human, who has most of the muscles that move the fingers located in the forearms, with very long tendons extending into the hands.

Oster, G. "Muscle Sounds." Scientific American *255 (1984): 108.*

Oster, G., and J. S. Jaffe. "Low-Frequency Sounds from Sustained Contraction of Human Skeletal Muscle." Biophysical Journal *30 (1980): 119.*

142. Sound in a Tube

A sound wave (or any wave) is partly reflected, partly transmitted, and partly absorbed whenever the wave encounters a change in resistance to its movement. A sound wave will be reflected from a solid wall because the sudden increase in density produces a change in the resistance. Different materials cause different phase changes for the reflected wave compared to the original wave.

A sound wave moving inside the tube encountering the open end will be partially reflected. A compression region at the open end will expand outward, thus creating a deficit of pressure—a rarefaction. Surrounding air gets pushed into this region to build up a compression region moving back into the tube. One can envision the opposite effects when a rarefaction reaches the end of the tube.

Effectively, the dynamic length L' of the tube is longer by about one-third of the open end diameter D for each open end if one desires to use the simple formula relating resonant wavelength λ to physical tube length L. That is, instead of the formula $2L = n\lambda$, one should use $2L' = n\lambda$, where $L' = L + 2D/3$, and the resonance tones will have a slightly lower pitch than expected from the first formula.

Troke, R. W. "Tube-Cavity Resonance." Journal of the Acoustical Society of America *44 (1968): 684.*

143. Those Summer Nights

Sound travels faster in warm, dry air than in cooler dry air. In warm air the average molecular speeds are greater, so molecules get to their neighbors sooner, enabling the sound compression to move faster. In summer, when the air temperature is warmer than the water temperature, a temperature inversion situation exists: The air temperature for several meters above the water can be lower than the air temperature above that layer. The temperature inversion will act to reflect upward-moving sound energy back toward the water, the calm water surface will reflect the sound wave back upward, the temperature inversion will reflect the sound back downward, and so on. Therefore, much of the sound will travel within a thin sheet of air to a far distance across the water surface. The intensity of the sound heard at a distant location depends upon several factors, such as the frequency, the original intensity, and the effective reflection coefficient of the temperature inversion. The direction changes of the wavefronts are easy to visualize.

144. Cannon Fire

The acoustic phenomenon experienced near London has several possible explanations. The simplest may be that the upper winds blow in a direction opposite to the lower winds. A westerly wind below and an easterly wind above will prevent the sounds nearer to the ground from reaching points west of the source. Then when the sound reaches the upper winds, the sounds can reach those points farther west by deflecting downward far away from the source. However, this process cannot explain a ring of sound in all directions.

When the zone of silence completely surrounds the source at some radius, but the sound is heard at greater distances in all directions, a different explanation is required. In this case in London, one guesses that there was a temperature inversion high in the atmosphere. The firing of the cannon sends out a hemispherical wavefront that expands as it rises above the ground. If the air temperature decreases with height, as it usually does, the wave bends away from the ground. Enough sound is usually diffracted back to the surface, especially at lower frequencies, so that the cannon fire can be heard easily over a considerable area around the source. But as the wave travels upward, the diffracted sound has less chance to reach the ground because the distance is increasing, and beyond a certain radius around the source there will be a zone of silence.

When this sound wave reaches a height of 10 to 15 kilometers the air temperature stops decreasing and begins to increase slowly with height to a maximum at about 50 kilometers. The temperature increases in this zone because it is heated by the ultraviolet radiation from the sun. Much of the intense ultraviolet radiation is absorbed by the ozone layer, which protects us from having our skin burned to a crisp. (Some UV gets through; otherwise we could not get a tan.)

On meeting the warmer air, the sound wave bends away from it and travels back toward the ground. Only a small amount of the sound energy survives this long journey back to the ground because there is geometric spreading into space and absorption by the air. Sounds of large explosions and artillery fire can be heard if the atmospheric conditions are favorable.

145. Speaking into the Wind

The wind cannot blow the sound back, unless the wind speed reaches the speed of sound! The wind actually lifts the sound upward on the upwind side so that most of the sound energy goes over your head. In the first diagram the velocities of the sound and wind add vectorially. (The length of the wind velocity vector is exaggerated.)

On most days the air temperature decreases with elevation because the air is mostly heated by the ground, not directly by the sun. Sound waves curve away from warmer air, so the pattern of the sound rays emitted by a point source located off the ground appears as shown in the second diagram, assuming no wind. The darker areas on either side of the source represent sound shadows where very little sound is heard.

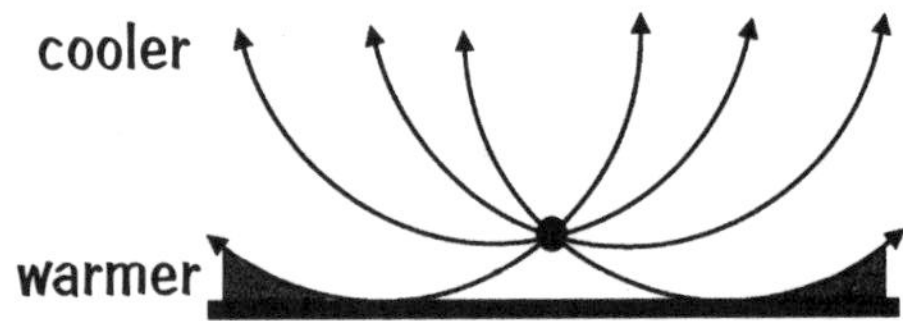

When there is a wind, its speed increases with elevation. Combine the sound speed and the wind speed vectorially, accounting for the increase in wind speed, to obtain the result shown in the third diagram. There is a sound shadow on the upwind side, but some sound diffracts into the shadow, particularly at the lower frequencies. Higher-frequency sounds, including those in speech, are effectively absent in the shadow region. Without the higher frequencies associated mainly with consonants, speech may be unintelligible. So the wind causes two effects: decreasing the intensity and removing the higher frequencies.

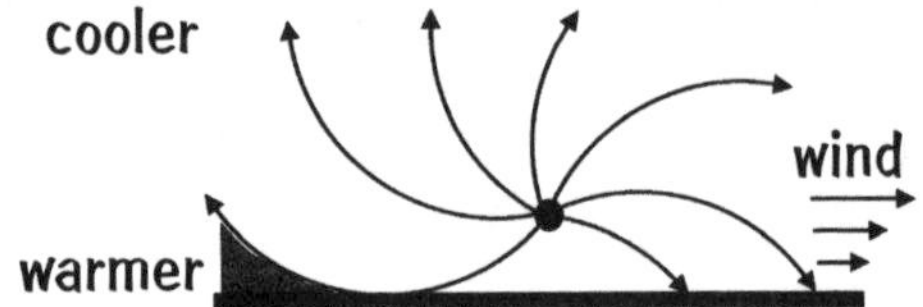

146. Foghorns

Low-pitched sounds can be heard at greater distances than higher-pitched sounds. As sound waves are transmitted, some of the energy is transformed into thermal energy, with the conversion rate being greater for the higher frequencies. Ships at sea need plenty of space to change course to avoid danger. Thus foghorns are always low-pitched to ensure that they can be heard across miles of water.

147. Yodeler's Delight!

In normal atmospheric conditions, the air temperature decreases with altitude. Therefore, the speed of sound in the air decreases with elevation. The sound waves originating near the ground spread outward from the source in all directions, eventually bending away from the warmer ground to move upward toward the balloonist or the mountain climber above.

Sound waves produced above by the mountain climber or balloonist start out going in all directions from the source, and they also bend away from the ground, often failing completely to reach the ground.

There are two other minor effects: (1) The balloonist or the mountain climber produces sounds in air of slightly lower density than on the ground, so the energy of the sound waves is less than the energy of the sound waves produced by people on the ground. (2) The balloonist is also in a region of quiet, while people on the ground are immersed in a flood of sound, making the balloonist's voice more difficult to pick out from the background noise.

148. Tuning Fork Crescendo

The two prongs produce sound waves of opposite phase. The two waves will practically cancel each other out when the prongs vibrate in a plane perpendicular to the plane of the ear. Rotated a quarter turn, the prongs vibrate in a plane parallel to the plane of the ear, and the two sound waves reinforce each other to produce a louder sound. Rotating the fork smoothly varies the intensity from loud to faint.

Crawford, F. S. Waves: Berkeley Physics Course. Vol. 3. *New York: McGraw-Hill, 1968, p. 532.*

Zarumba, N.; R. Hetzel; and E. Springer. Physics Teacher *21 (1983): 548.*

149. Hark!

No. Part of the sound emanating from the speaker reflects off the walls and the ceiling, and the rest is absorbed. A woman tends to emit higher-pitched tones, and these are absorbed to a greater extent than lower-pitched tones. Hence bass and tenor notes are reflected a greater number of times, so a male speaker needs to expend less energy to fill the room with his voice. However, he must speak more slowly to avoid beginning his next word simultaneously with the end of his last word, which may still be flying around the room!

150. Rubber and Lead

The velocity of sound in a material depends upon both the density *and* the elasticity—sound velocity = $\sqrt{\text{elasticity/density}}$. Lead has a very low elasticity value—it does not spring back well. Cooling lead can improve its elasticity tremendously. Rubber is another exception because of its extreme sponginess and peculiar chemical structure, both features allowing the sound energy to be absorbed readily.

Since both materials have low sound velocities and are unable to transmit sound energy efficiently, a sandwich of alternating layers of lead and rubber is used to isolate equipment from floor vibrations in many research labs.

151. Helium Speech

The frequencies of the vocal folds in the human windpipe are independent of the gas surrounding them and are determined by their mass and tension. While their frequency spectrum remains the same when breathing helium, the associated speech/mouth cavity enhances harmonics selectively by resonance, changing the intensities without changing the frequencies themselves. The result is similar to turning up the treble on your stereo.

A helium atom is less massive than any type of molecule in air except the trace molecule of hydrogen, H_2, and the speed of sound in helium gas is greater than in air. For a sound wave, the frequency = velocity/wavelength. Therefore, sounds of the *same wavelength* also have a higher frequency in helium than in air.

Tibbs, K. W., et al. "Helium High Pitch." Physics Teacher *27 (1989): 230.*

Van Wyk, S. "Acoustics Problems." Physics Teacher *25 (1987): 521–522.*

152. Maestro, Music Please!

The first design, with the curved reflector, has better acoustics. If the listener hears the first reflected sound fewer than 50 milliseconds after the direct sound, the reflected sound will tend to reinforce the direct sound, and the effect will be pleasing. If the delay is 50 milliseconds or longer, the listener will hear the reflection as an echo, which interferes with the direct sound. Multiple reflections are less important due to absorption of sound energy.

Blum, H. American Journal of Physics *42 (1974): 413.*

Rossing, T. D. "Acoustic Demonstrations in Lecture Halls: A Note of Caution." American Journal of Physics *44 (1976): 1220.*

153. The Mouse That Roared

Although a mouse may be able to generate low-frequency sounds in its mouth cavity, their intensity is very limited by two factors: the small amount of air being moved within the mouth, and the great mismatch in size between the wavelength of the sound and the largest linear dimension of the mouse's oral cavity. Cavity resonance effects would be almost nonexistent; they cannot help the poor fellow. And the intensity of sound, with all other parameters the same, depends upon the square of the frequency, so one must move a significantly greater volume of air at the lower frequencies to match the intensities at higher frequencies. The mouse cannot move a large volume of air!

The elephant can emit high-frequency sounds by several mechanisms, including via small-size resonant cavity contributions in the mouth or nose as well as through the ability to utilize nonlinear vibrational behavior, which introduces higher-frequency harmonics.

Bartlett, A. A. "The Mouse That Roared?" Physics Teacher *15 (1977): 319.*

154. Bass Notes Galore

Human speech consists of both the lowest fundamental tones and their harmonics—whole-number multiples of these fundamental frequencies. The human ear-brain system not only senses the frequencies present in a sound wave but also produces new frequencies, which are the sums and differences of those originally present. This capability arises in most systems that exhibit nonlinear responses to input signals. The bass tones heard in the sound from a telephone speaker arise from the difference frequencies.

Rossing, T. D. "Physics and Psychophysics of High-Fidelity Sound." Physics Teacher *17 (1979): 563–570.*

Stickney, S. E., and T. J. Englert. Physics Teacher *13 (1975): 518.*

155. Virtual Pitch

When two tones are sounded together, a third, lower tone is often heard. This undertone is called a difference tone or Tartini tone, after the Italian violinist who described it in 1714. If the two original tones have frequencies f_1 and f_2, this difference tone is at $(f_2 - f_1)$. One can also hear the cubic difference tone at $2f_2 - f_1$, and possibly others with difficulty. These difference tones rely upon the nonlinear response of the human ear-brain system where a quadratic response term adds to the linear response term. Tibetan monks sometimes sing choral music that contains voices at 600 hertz, 800 hertz, 1,000 hertz, and 1,200 hertz, for example, and one hears many of the difference tones.

Hall, D. E. "The Difference between Difference Tones and Rapid Beats." American Journal of Physics *49 (1981): 632–636.*

156. Singing in the Shower

Good singing requires resonances. The sound originates with the passage of air pushed out by the lungs through the vocal folds (membranes that are often miscalled vocal cords) of the human windpipe as a series of air pulses with a frequency determined by the tension of the vocal folds. The sound is a harmonic series of sound waves encompassing the fundamental frequency and the higher harmonics, with the fundamental being the most intense. As these sound waves pass through the vocal tract consisting of the larynx, the pharynx, and the mouth, those frequencies close to the resonant frequencies of the vocal tract will be louder than the others. A good singer can achieve the match in several ways, by adjusting the tension in the vocal folds and by varying the shape of the vocal tract, thereby benefiting from the amplification arising from the resonance. Without the aid of the resonances, one might need to scream to be heard by the audience at those frequencies!

The advantage of singing in the shower is that the unskilled vocalist now has the help of the resonances generated between the surfaces of the shower enclosure. A closed shower essentially has three resonance directions: (1) between the floor and the ceiling, (2) between the front and the back walls, and (3) between the two sidewalls (treating the shower door or curtain as a wall). A standing wave of sound resonance can be established between any pair of walls, with pressure antinodes at the walls and a node at the center for the fundamental frequency in that direction. The second harmonic at *twice* the fundamental frequency has three antinodes and two nodes. From the relation frequency = velocity/wavelength, one can predict some of the resonant frequencies, knowing that the fundamental's wavelength will be about twice the distance between reflecting surfaces. For a floor-to-ceiling distance of 2 meters, for example, the fundamental's wavelength is 4 meters with a frequency of 86.5 hertz, taking the speed of sound to be 346 meters per second.

To excite the fundamental, one cannot stand where the node is supposed to be—near the center. One must stand nearer to one wall—that is, nearer to the antinode. The second harmonic and all the other even harmonics can be excited from the center region. How good the sound seems depends upon several factors, including the location of the ears and the mouth (the sound source), and the distortions of the sound by the head and the body. The shower singer needs to move around until pleasing effects occur. Usually the third, fourth, seventh, and eighth harmonics come out best for a person standing in the shower. Of course, all three directions must be considered simultaneously, for higher harmonics may resonate in one of the other directions. May your singing sound great!

Edge, R. D. "Physics in the Bathtub – or, Why Does a Bass Sound Better while Bathing?" Physics Teacher *23 (1985): 440.*

Walker, J. "What Makes You Sound So Good when You Sing in the Shower?" Scientific American *253 (1982): 170–177.*

157. Scratching Wood

The sound is very soft when you listen to it in the air because the sound energy spreads out in all directions and the geometrical factors dictate that a very small fraction reaches your ears. You hear a louder sound with your ear right up against the piece of wood because the scratching produces sound in the wood as well as in the surrounding air. Most of the sound energy in the wood remains in the wood because there is a large impedance mismatch at the wood-air interface to efficiently reflect most of the sound energy and permit very little sound to transfer to the air. So your ear receives more sound energy from the wood if the contact is good.

158. Simple String Telephone Line

Line *B*, with the cup reversed from the traditional way. This orientation places the vibrating surface, the cup bottom, closer to the ear, which produces a louder sound. Give it a try. One now wonders whether the sending cup should also be reversed!

Heller, P. "Drinking-Cup Loudspeaker—A Surprise Demo." Physics Teacher *35 (1997): 334.*

159. Supersonic Aircraft

When a plane is moving subsonically, its sound waves precede the plane, causing the air molecules in front of it to spread out in nonconcentric spheres spaced more closely in the forward direction than in the backward direction.

When a plane is moving at supersonic speeds, the air molecules receive no advance warning. In fact, shock waves are created at many leading edges on the plane, all of which tend to coalesce into two apparent source locations, one near the plane's bow and one near the tail. Consequently the supersonic plane experiences more turbulence, greater drag forces, and more heating along the leading edges. Particular wing configurations reduce the vibrations, and special metals and exotic materials are used that are better able to tolerate the higher temperatures.

As the two shock waves travel downward to the observer on the ground, the first shock wave, from the bow of the plane, raises the air pressure sharply. Then the air pressure decreases to below atmospheric pressure with the advent of the tail shock wave, and then rises sharply again. Hence the two booms, one at each sharp rise of the pressure.

Hodges, L. "What Are the Effects of a Sonic Boom?" Physics Teacher *23 (1985): 169.*

160. Slinky Sound Off!

Radiating from the output at the wall will be a "whistler," a sound that first becomes audible as a very high pitch, then quickly descends in pitch, becoming inaudible in a fraction of a second. The Slinky under very little tension behaves like a stiff long bar, and the speed of the sound waves will be proportional to the square root of the frequency. Thus, higher-frequency sound waves travel faster than the lower-frequency ones.

Crawford, F. S. "Slinky Whistlers." American Journal of Physics *55 (1987): 130.*

161. Wineglass Singing I

They sound slightly different. Rubbing the glass mainly excites the lowest "bell mode," the (2,0) mode, with two nodal meridians. Tapping the glass excites many more of these "bell modes," including the (2,0) and the (3,0) modes.

Rossing, T. D. "Wine Glasses, Bell Modes, and Lord Rayleigh." Physics Teacher *28 (1990): 582.*

162. Wineglass Singing II

Go ahead and do it! The frequency of the sound decreases even though the air column is getting shorter. The vibrations of the glass wall must move more mass, including itself and the added water, increasing the inertia.

Rossing, T. D. "Wine Glasses, Bell Modes, and Lord Rayleigh." Physics Teacher *28 (1990): 582.*

163. Bell-Ringing Basics

Unlike most string and pipe instruments, bells have overtones that are not harmonics—that is, that are not integer multiples of the fundamental frequency. These overtones produce unpleasant beats either among themselves or with one of the fundamentals.

164. Forest Echoes

For the echo to be raised an octave, the wavelength of the original sound must be greater than the spacing of the trees, which are the scattering centers. Under this condition, Rayleigh scattering (i.e., coherent scattering) of the sound waves will occur, and the scattering intensity is proportional to the *fourth* power of the frequency. Thus the harmonic at twice the fundamental frequency is returned at sixteen times its original intensity and may dominate the returning sound!

Rayleigh, Lord. Nature *8 (1873): 319.*

Rinard, P. M. "Rayleigh, Echoes, Chirps, and Culverts." American Journal of Physics *40, 923 (1972): 923.*

165. Bass Boost

The sensitivity of the human ear varies with the frequency and the quality of the sound. Fletcher and Munson determined curves of equal loudness many years ago, and their measurements demonstrate the relative insensitivity of the human ear to sounds of low frequency at moderate to low intensity levels. Hearing sensitivity reaches a peak between 3,000 and 5,000 hertz, which is near the resonant frequency of the outer ear canal. So when the stereo level is turned down, the bass must be turned up.

Fletcher, H., and W. A. Munson. "Loudness, Definition, Measurement, and Calculation." Journal of the Acoustical Society of America *6 (1933): 59.*

Rossing, T. D. "Physics and Psychophysics of High-Fidelity Sound." Physics Teacher *17 (1979): 563–570.*

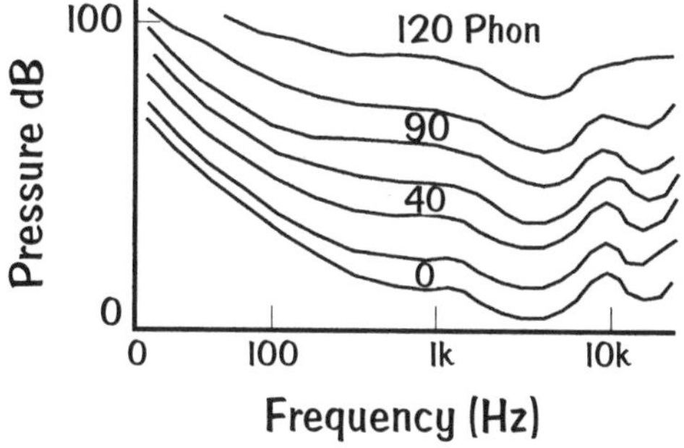

166. Personal Attention-Getter

An array of several small speakers, all located within a one-meter or less radius, can be used if a high-frequency audio carrier is used to transport the low-frequency voice message. The array can be designed to take advantage of the phase relationships of the several speakers to send a focused beam to the desired recipient in the crowd. The minimum focus diameter at the recipient will be the carrier wavelength, as dictated by wave dynamics.

167. Musical Staircase

The human mind tends to form linkages between elements that are close together rather than those that are far apart. For example, human vision helps us group dots that are next to one another, like the image we see in the television screen. Our vision also alerts us to be more sensitive to neighboring lights turning on and off than for images that are farther apart. Likewise, human sound perception behaves so we prefer to recognize notes of the musical scale that are closer together rather than notes that are farther apart. Hearing research has indicated that the twelve notes of one octave are often perceived as existing in a circle called the pitch class circle. Among the examples investigated, the playing of two sets of three notes of the octave pitch class circle in sequence will be heard differently by different listeners. If you start by playing D and B simultaneously, followed by E and A together, then F and G together, some listeners will hear the sequence BAG at a higher pitch than DEF, while others will hear BAG as lower notes than DEF.

But what an individual hears also depends upon the language or dialect spoken by that person. For the latest details of this ongoing investigation, begin with the article below.

Deutsch, D. "Paradoxes of Musical Pitch." Scientific American *263 (1992): 88–95.*

168. Where Does the Energy Go?

When two acoustic waves cancel by destructive interference in a region, one can add the two wave amplitudes to get zero amplitude. But the power carried by the sound wave, which is a product of field intensities and the acoustic wave impedance, *cannot* be determined by superposition.

If one uses two closely spaced loudspeakers, one may drive them out of phase to produce almost total cancellation of their acoustic radiation. Electrical energy is still going into both speakers—one only needs to measure the currents driving the speakers. The reason for less radiation lies with the acoustic impedance of the air, a derived quantity that varies with the output of the other acoustic sources in the environment. For two identical out-of-phase speakers, the *true acoustic impedance has been reduced to zero.* The power is calculated from this relation: power = wave amplitude squared times the acoustic impedance. The power is now equal to zero watts. In other words, the impedance mismatch results in no energy being radiated into the air. Let Z_1 be the acoustic impedance of the air and Z_2 of the speaker. If the ratio of acoustic impedances $Z_1/Z_2 = 1$, all the energy is transmitted and none is reflected. In our case, $Z_1/Z_2 = 0$.

Levine, R. C. "False Paradoxes of Superposition in Electric and Acoustic Waves." American Journal of Physics *48 (1980): 28–31.*

*169. A Bell Ringing in a Bell Jar

Although one at first might think that the demonstration shows the inability of sound to be transmitted through a gas at low pressures, what really happens is a very inefficient transfer of sound energy from the vibrating bell into the air at reduced pressure because there exists a tremendous acoustic impedance mismatch. (Acoustic impedance is the resistance to the flow of acoustic energy.) Sound travels well through a gas as long as the sound wavelength is large compared with the mean free path for the air molecules. Even at 1,000 N/m^2 (10^{-2} atmosphere), the mean free path is about 10^{-3} cm, far less than the approximately 10 cm wavelength of the sound from the bell.

So the real problem is that less and less acoustic energy is being transmitted from the bell into the air, and from the air to the glass of the bell jar. How much sound energy is transmitted and how much is reflected depend upon the acoustic impedances of the two media. The transmitted amount depends upon the ratio Z_1/Z_2 of the acoustic impedances, where $Z = \rho v$, with ρ being the density of the medium and v the velocity of sound in the medium. When $Z_1/Z_2 = 1$, all the sound is transmitted and none is reflected. Even at atmospheric pressure the impedance of the air is much less than for glass or metal, and the ratio becomes smaller and smaller as the pressure is reduced.

Chambers, R. G. Physics Teacher *9 (1971): 272, 369.*

*170. A Well-Tuned Piano

Western music is based on scales defined by certain frequency ratios of integers between successive notes. In the so-called natural or ideal system, going all the way back to Pythagoras, the ratios within an octave are:

C	D	E	F	G	A	B	C
1.000	1.125	1.250	1.333	1.500	1.667	1.875	2.000
24/24	27/24	30/24	32/24	36/24	40/24	45/24	48/24

This scale can be extended upward into the next octave by simply doubling all the numbers, or downward, by halving them. A piano tuner could adjust all the white keys on a piano to this sequence of pitches, and you could play many different kinds of simple music.

Suppose you decide to play a simple melody that normally began on C in a new way, by starting on the next note of the scale—the note D. The result would be odd because the tune now played would not sound like the original tune. The discrepancy would be even greater if we began on a note farther from C. A satisfactory solution to this problem was found by introducing the even-tempered system more than 250 years ago, and now you can play any melody equally well starting from any note.

In the even-tempered scale the octave is divided into twelve equal semitone intervals, so that any two successive semitones have the same frequency ratio. Since each note must vibrate at twice the frequency of the same note one octave below, the semitone ratio from note to note is taken as the twelfth root of 2, namely, 1.05946. This solution gives a continuous geo-

	C	C#	D	D#	E	F	F#	G	G#	A	A#	B	C
Ratio	1.000	1.0595	1.1225	1.1892	1.2600	1.3348	1.4142	1.4983	1.5874	1.6818	1.7818	1.8877	2.0000
Frequency	261.63	277.18	293.66	311.13	329.63	349.23	369.99	391.99	415.31	440.00	466.16	493.88	523.25
Ideal Scale Ratio	1.000		1.1250		1.2500	1.333		1.5000		1.6666		1.8750	2.0000

metric progression throughout the keyboard, and the C scale is approximated by the frequencies (in hertz) shown in the table and the nearly same ratios.

On a piano keyboard, the tuner does not make any difference in tuning between black and white keys—all are arranged in a uniformly rising sequence of pitch. The two colors and shapes of keys are there only to help the player find her way by feel over the wide expanse of the keyboard.

Ultimately, with the even-tempered system the sequence of pitch is not a precise agreement with the natural scale, but it does provide a close approximation. In fact, the modern ear (since the time of Bach in the 1700s) has become so accustomed to the "errors" that this tuning scheme sounds correct!

Bernstein, A. D. "Tuning the Ill-Tempered Clavier." American Journal of Physics *46 (1978): 792–795.*

*171. Driving Tent Stakes into the Ground

The very different behavior is explained by the mismatch of the acoustic impedance of each material with the acoustic impedance of the soil, where the acoustic impedance $Z = \rho v$, with ρ being the density of the medium and v the velocity of sound in the medium. The blow of the hammer sets up a transient stress wave in the stake, and when it reaches the end of the stake that is in contact with the ground, part of the wave is reflected and part is transmitted into the ground. If the ratio of acoustic impedances $Z_1/Z_2 = 1$, all the energy is transmitted and none is reflected. This transmitted wave tends to break up the soil.

For steel, the mismatch is much greater than for wood, so most of the wave in steel will be reflected at the junction and most of the momentum imparted by the hammer will remain in the stake. The steel stake will acquire a high velocity and move into the dirt.

Rinehart, J. S. "On the Driving of Tent Stakes." American Journal of Physics *19 (1951): 562.*

———. *"A Demonstration of Specific Acoustic Resistance."* American Journal of Physics *18 (1950): 546.*

*172. Loudness

Doubling the sound intensity level does not usually make the perceived sound doubly loud for the ear-brain system because the human loudness response does not follow the traditional logarithmic dB scale. For different frequency ranges of sound, one measures different responses to loudness change. Usually one needs an increase in the sound intensity level of between 6 and 10 to hear double the loudness—that is, the subjective perception is much different from a sound intensity meter response, which simply senses sound pressure. Recent meters have begun incorporating this different human response for different frequencies into their design, so meters are now available that correspond very well to the human response curves.

Even without considering the human ear-brain response, one knows that lower-frequency sound will need much larger amplitude sound waves to deliver the same amount of sound energy because the energy/second in any wave is proportional to f^2A^2, where f is the frequency and A is the amplitude. Simply doubling the frequency means the speaker excursion distance can be half as much for the same energy/second emitted if the acoustic impedance of the medium is the same.

Rossing, T. D. "Physics and Psychophysics of High-Fidelity Sound." Physics Teacher *17 (1979): 563–570.*

Chapter 6 Opposites Attract

173. Three-Bulb Circuit

The voltage across lamp 3 becomes zero, so it no longer glows. Lamps 1 and 2 glow brighter than before because the battery voltage is now shared equally by two identical lamps instead of three.

Hewitt, P. "Figuring Physics." Physics Teacher *26 (1988): 313–314.*

174. Potato Battery

The small flashlight bulb does not glow perceptibly. The potato battery has enough terminal voltage but is unable to deliver more than a few microamperes of current at that voltage. One can run an LCD clock with the potato battery because this clock demands only microamps of current.

Stankevitz, J., and R. Coleman. "A Curious Clock." Physics Teacher *23 (1985): 242.*

175. Resistor Networks

The total resistance of each circuit is the same. Therefore they require the same current values.

Feynman, R. P.; R. B. Leighton; and M. Sands. The Feynman Lectures on Physics. *Reading, Mass.: Addison-Wesley, 1964, page 22–12.*

176. A Real Capacitor

Only an isolated ideal capacitor could keep its electrical charge forever. A real capacitor has an effective resistance value across its plates. As an example, some small 5-*V*, 1-Farad capacitors have a discharge time of about five seconds or so. This value is their *RC* time constant *T*, so that the internal resistance value $R = T/C$, or $R = 5$ ohms. Most capacitors have a much longer RC time constant.

French, A. P. *"Are the Textbook Writers Wrong about Capacitors?"* Physics Teacher *31 (1993): 156–159.*

Kowalski, L. *"A Myth about Capacitors in Series."* Physics Teacher *26 (1988): 286–287.*

177. Capacitor Paradox

Suppose the two capacitors each have capacity *C*, and that capacitor *A* is charged to a voltage $V = CQ$. The energy in capacitor *A* is then $1/2\ CV^2$. When the capacitors are connected, the charge is shared equally, so the voltage drops to $1/2\ V$. The total energy in the two capacitors is now $C(V/2)^2$, which equals $1/4\ CV^2$. This difference in total energy in the capacitors has been the source of the thermal energy that heated the resistance wire.

If $R = 0$, the current in the wire establishes a magnetic field. If the current value oscillates, the changing current radiates electromagnetic waves, so the energy is radiated away with elapsed time.

Powell, R. A. "Two-Capacitor Problem: A More Realistic View." American Journal of Physics *47 (1979): 460–462.*

178. Charge Shielding

Yes, connect the metal shield to a good ground such as the earth. Before the grounding, there were equal and opposite charges on the inside and the outside surfaces of the metal shield. The ground connection lets the charges on the outside of the metal shield spread over a larger surface provided by the ground, so the amount of outside charge approaches zero. The inside charge on the metal shield is held there by the initial opposite charge to be shielded. By Gauss's law applied around the shield, the total charge inside the metal shield is zero.

179. Three Spheres

One might guess that the three spheres have the same charges. If the three spheres were at the corners of an equilateral triangle with three wires connecting the three pairs, then that guess would be correct.

However, the arrangement of the spheres has bilateral symmetry about the center sphere, so the two end spheres would have the same charge value—say, q. Let the center sphere have charge q'. The electrical potential V at the center of a sphere is the charge value q divided by the distance r to the charge, or $V = q/r$. For an isolated charged sphere of radius R and charge q, the potential $V = q/R$.

For our three spheres, the potential at the center of the middle sphere is $V = (2q/50 \text{ cm}) + (q'/10 \text{ cm})$. The potential at the center of each end sphere is $V = (q/10 \text{ cm}) + (q'/50 \text{ cm}) + (q/100 \text{ cm})$. When you solve these equations you get $q = 8Q/23$ and $q' = 7Q/23$. This charge distribution maintains the same constant potential on all three spheres.

180. Inductive Charges?

One can charge the electroscope by induction, meaning that the initially charged object does not transfer any of its charge to the electroscope because these two objects never contact each other.

Bring the negatively charged rod near the top of the electroscope. The leaves of the electroscope will separate to indicate that they now have like charges that repel. In fact, the leaves have excess negative charges (repelled from the top by the negative rod nearby), and the top has an excess of positive charge. Be sure to keep the negatively charged rod in place near the top of the electroscope while bringing one fingertip near the edge of the electroscope top. A small spark can be heard and the leaves collapse, indicating that the leaves no longer have an excess charge but have become neutral. Remove the finger and then remove the charged rod. The electroscope leaves are now separated in this final condition.

Little Stinkers. "Charging of an Electroscope." Physics Teacher *3 (1965): 185.*

181. Parallel Currents I

Yes. The positive residues of the atoms move in the other direction to produce identical currents and their magnetic fields.

182. Parallel Currents II

In the frame S', the total force has two contributions: the attraction from the two parallel currents, and the repulsion from the electric force. Unlike the case of two parallel wires carrying currents, where there are as many negative charges as positive charges in the wire, the opposite charges do not exist for the case presented here. The repulsive electrical force is always stronger than the attractive magnetic force until the speed reaches the velocity of light, an impossible achievement.

Tilley, D. E. "A Question on Charge Interaction." Physics Teacher *14 (1976): 115.*

183. Rotating Wheel

The electric field due to Q at any point distance is exactly the same, in the air and in the oil. The result: no torque.

Chambers, R. G. Physical Education *12 (1977): 212, 229.*

184. Charge Trajectory

No. The electric force will be tangential to the electric field line, but there is no centripetal force. The test charge trajectory cannot curve along the field line.

Kristjansson, L. "On the Drawing of Lines of Force and Equipotentials." Physics Teacher *23 (1985): 202.*

Sandin, T. R. "Viscosity Won't Curve It." Physics Teacher *24 (1986): 70.*

185. Voltmeter Reading

The voltage across the 12-volt battery is 12 volts *whether or not* there is a net current through the 4-ohm resistor. The voltmeter will read 12 volts.

Viens, R. E. "A Kirchoff's Rules Puzzler." Physics Teacher *19 (1981): 45.*

186. Power Transfer Enigma

Maximum power transfer occurs when $R = r$, when the efficiency of power transfer is only 50 percent. As R grows bigger than r, the amount of power transfer decreases to zero at 100 percent efficiency.

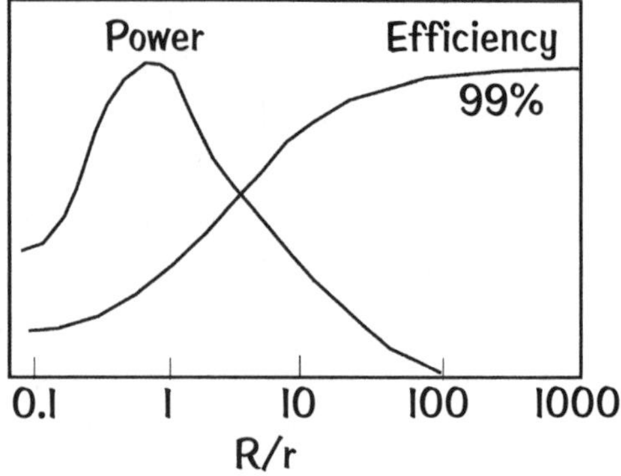

Hmurcik, L. V., and J. P. Micinillo. "Contrasts between Maximum Power Transfer and Maximum Efficiency." Physics Teacher *24 (1986): 493–494.*

Kaeck, J. A. "Power Transfer in Physical Systems." Physics Teacher *28 (1990): 214–221.*

187. Linear Resistance

No. The standard resistor behaves linearly only when its power dissipation is within its power rating—that is, when operating within its designed temperature range. Overheating the resistor makes its behavior unpredictable with a nonlinear response.

188. Radioactive Currents

Zero. This spherically symmetric current distribution has lines of current radiating outward from the center, but the magnetic field from each ray of current is canceled by the fields from the other rays. Otherwise the source would act as a magnetic monopole, which we know does not exist.

Brain Teaser. Physics Teacher *9 (1971): 405, 434.*

189. Which Is the Magnet?

Place the two bars as shown so that they make a *T*. If the top of the *T* is the permanent magnet, there will be no attraction between them.

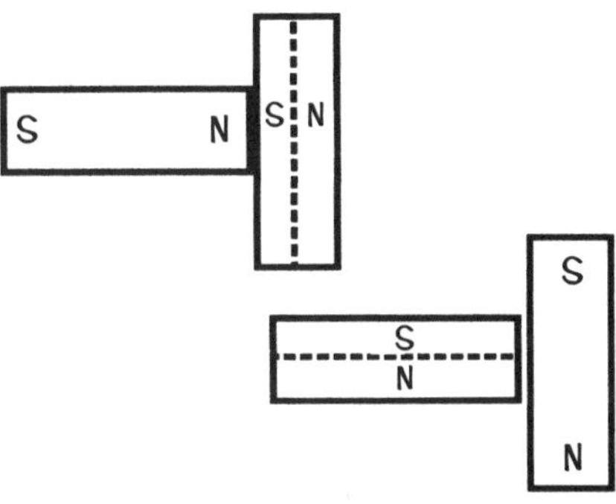

190. Why the Keeper?

Without a keeper, many of the magnetic field lines from the north pole to the south pole of the magnet bulge out into the surrounding space. Following these lines into the material at both poles reveals that their directions are not along the desired permanent magnet field line directions that establish the strong poles. A thermal or mechanical shock would lead to a possible disalignment of the magnetic domains, which would then establish slightly different directions for their minimum energy states. One can prevent this weaker misdirected magnetization from occurring by using a keeper that ensures that practically all the magnetic field lines between the poles are properly directed.

191. The Magnet

When bar *B* is placed on the magnet, some of the magnetic field lines are "short-circuited" through bar *B*, reducing the number of magnetic field lines through bar *A*. The attractive force between bar *A* and the magnet is significantly reduced, and the bar falls off.

192. Magnetic Sphere

If the sphere were put together as described, the sphere would be observed to have no magnetic properties because it would have become demagnetized during assembly. The sphere is symmetric under all rotations, so that if any point of the sphere has a field line in a particular direction, a 180-degree rotation about an axis connecting that point and the sphere center should bring back the original state. The rotation fails to do so, unless an opposite magnetic field line also exists through the point. But these two equal and oppositely directed field lines add to zero net magnetic field. QED.

193. Two Compasses

The two compasses will behave like two weakly coupled oscillators. The oscillation of the second needle will decrease as the first one oscillates with increasing angular displacement. Then the energy transfer goes the other way. Eventually friction damps out the oscillations.

One can observe the two normal modes of oscillation by shaking both compasses initially. Many complex modes of behavior for coupled oscillators can be made visible with this system.

Snider, J. L. "Simple Demonstrations of Coupled Oscillations." American Journal of Physics *56 (1988): 200.*

194. Magnetic Work?

Imagine a magnified piece of wire placed in a magnetic field that points into the page (see the diagram). Suppose a current flows in the wire toward the top of the page. There will be a sideways force $F = -ev \times B$ on the electrons constituting the current. As a result, the electrons will tend to drift toward the right. An excess of electrons on the right and a deficit on the left will create a repulsive force on the electrons drifting toward the right. This is known as the Hall effect. The electrons will keep piling up on the right until the repulsive force becomes strong enough to counterbalance the force due to the magnetic field and there is no longer a net force on the electrons. Note, however, that the positive ions of the metal have fixed positions and no magnetic force acts on them. But they are very much subject to the electric force due to the pileup of the electrons on the right. This electric force pulls the ions to the right, thus producing the motion of the wire as a whole. The paradox is therefore resolved by observing that the motion of the wire is caused by an electric field, not a magnetic field. Note also that one cannot consider the wire as a closed system because charges continue to enter one end of the wire and leave by the other.

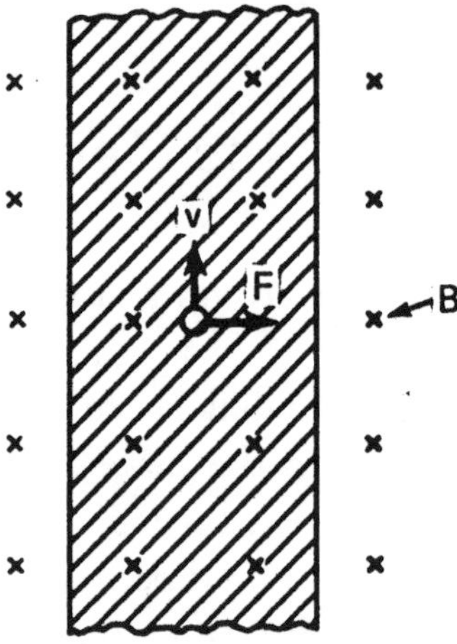

Coombes, C. A. "Work Done on Charged Particles in Magnetic Fields." American Journal of Physics *47 (1979): 915–917.*

Mosca, E. P. "Magnetic Forces Doing Work?" American Journal of Physics *42 (1974): 295–297.*

195. Electric Shield

Yes. Without an electric field, the magnetic-field part of an electromagnetic wave cannot propagate. Therefore a Faraday cage, an enclosure made of metal screen, can prevent electromagnetic waves from propagating into the cage as long as the wire spacings are smaller than the wavelength and the thickness is greater than the electromagnetic skin depth.

196. Wave Cancellation in Free Space

As far as the total electric field or the total magnetic field for a combination of waves is concerned, adding the fields of each separate wave together is a valid procedure. But the power (i.e., energy/time) is actually the product of the field intensity and the wave impedance. If the wave impedance were independent of other waves being present, then one could simply use the addition-of-the-fields analog. However, the impedance depends upon which other fields are present, so superposition for the power is not a correct procedure.

Levine, R. C. "False Paradoxes of Superposition in Electric and Acoustic Waves." American Journal of Physics *48 (1980): 28–31.*

197. Repulsion Coil I

The EMF around the string loop will be the same as the EMF around the metal ring. However, the string loop will not be repelled because there will be no induced current in the string and therefore no induced magnetic field. Therefore the string will not levitate.

198. Repulsion Coil II

Essentially the metal ring becomes the equivalent of a bar magnet with its poles facing the direction opposite to the poles of the repulsion coil itself. The upward magnetic repulsive force must be greater than the downward gravitational force for the jump to occur. One must account for the 180-degree phase change for the system: 90 degrees for the Faraday induction law plus 90 degrees for the inductance of the ideal ring, assuming that the ring has no electrical resistance.

Mak, S. Y., and K. Young. "Floating Metal Ring in an Alternating Magnetic Field." American Journal of Physics *54 (1986): 808–811.*

199. Magnetic Tape

The audiotape is a good electrical conductor, so the electrical charges spread out uniformly around the whole tape. The minimum energy shape would be a circle. One can demonstrate this response by charging the tape and suspending it in the air over an electrically charged PVC pipe.

200. Kelvin Water Dropper

Initially there will be a charge asymmetry, ever so slight, due to cosmic rays, natural radioactivity, etc. Assume can A is slightly negatively charged with respect to can B. The water in the nozzles responds to this difference in the top cans, letting positive drops fall through can A into the positively charged can C against the force of electric repulsion. Can C becomes more positively charged than before. The same process is happening on the other side, with can D becoming more negative. One can see that the charged droplets repel each other and break up into a spray of smaller droplets approaching the bottom cans. One can even see visible sparks when a sudden discharge occurs.

*201. Back EMF

No. The back EMF is the energy per charge that *drives* the motor to become mechanical work.

The potential difference V across the motor equals the sum of the back EMF value E and the IR drop associated with the heat generated. If we simplify the situation by ignoring the friction of the motor and the magnetic hysteresis, assuming that the electrical resistance R is temperature independent, etc., then E represents the mechanical energy output per unit charge. As the motor starts up, $E = 0$, and the current I is limited by the electrical resistance of the armature. If there is no mechanical load, the back EMF energy is converted into the mechanical kinetic energy of the rotor as it speeds up. With the increase in rotational speed, E approaches V, and I goes toward zero. In the limit, there is no current as the motor spins and no energy is converted.

With a load the motor slows, so E decreases to let I increase. The mechanical power delivered to the load is EI.

Lehrman, R. "The Back emf of a Motor." Physics Teacher *21 (1983): 315.*

*202. Axial Symmetry

If the two electrodes were parallel flat plates, the electric field would be uniform between the plates. We would expect the neutral particle to experience equal tugs in both directions no matter where the neutral particle is located between them.

In the axial symmetry case, the electric field is much stronger near the central charged wire. The neutral particle responds by accelerating inward toward the wire. The electrical force value is directly proportional to the polarizability of the neutral particle and to the gradient of the electric field—that is, how nonuniform the field is. This resulting attractive effect is called dielectrophoresis. We have here the electrical analog of the magnetic moment in a magnetic field gradient.

One can use this effect to separate out particles with different polarizability, such as powders in liquids. Large electric-field gradients can be built up easily because the effect works just as well for AC fields as for DC fields.

Pohl, H. A. "Nonuniform Electric Fields." Scientific American *239 (1960): 107–116.*

*203. A Ring Is a Ring . . . !

Measurement with a standard voltmeter would show no voltage. A standard voltmeter measures a scalar potential difference, while the EMF around the ring represents the vector potential.

There have been many articles in the scientific literature about the voltage to be measured, and they all are concerned with the geometry of the wire leads from the voltmeter and how much flux passes through the loops formed by these leads and the two sections of the ring.

The following argument justifies the zero-voltage reading. If one uses a twisted-wire configuration for the voltmeter leads to eliminate any loop contribution beyond the ring itself, lead connections at opposite ends of the diameter of the ring produce a symmetrical situation. There can be no voltage difference in this situation, for the EMF directions are the same around both loops, with a zero net value across the common diameter with its opposite current directions.

Varney, R. N. "Electromotive Force: Volta's Forgotten Concept." American Journal of Physics *48 (1980): 405–408.*

*204. Electromagnetic Field Energy

The article referenced below indicates that perhaps this question remains in the realm of current research questions. In contrast, one can apply the Heisenberg Uncertainty Principle to handle this question to determine where the energy in the electromagnetic field is when both the electric field and the magnetic field simultaneously go to zero.

On the small scale, the quantum mechanical effects are more dramatic than on the large scale. To find out where the energy is in the small region where the classical electromagnetic wave goes to zero, you must use the rules of quantum mechanics.

The Heisenberg Uncertainty Principle states that the uncertainty in the position Δx times the uncertainty in the momentum Δp must be greater than or equal to Planck's constant h. To identify the location in the electromagnetic wave where the fields simultaneously reach zero, you would make Δx smaller, so Δp grows larger. But $\Delta p = \Delta$ energy/c for an electromagnetic wave, so the uncertainty in the value of the energy increases. This energy uncertainty in the measurement will be large enough to accommodate the original energy in the classical electromagnetic wave.

Bueche, F. J. "Where's the Energy?" Physics Teacher *21 (1983): 52.*

*205. Levitating Top

This remarkable toy, called the Levitron, levitates in air a 22-gram spinning permanent magnet (ceramic) top, which floats about 3 centimeters above the magnetic base plate until its spin declines to fewer than about 1,000 revolutions per minute. Vertically, in equilibrium, the upward magnetic repulsion force between the two permanent magnets equals the downward gravitational force—that is, the weight of the top.

The spinning top has angular momentum about a nearly vertical axis. If the top tips slightly, it begins to precess instead of flipping over when the spin rate is above about 1,000 revolutions per minute. Too great a spin rate also causes problems!

Horizontal drift is limited by curving the base plate so that its magnetic field has a gradient in the region of the top, and the restoring force is enough to push the top back toward the center.

Berry, M. V. "The Levitron: An Adiabatic Trap for Spins." Proceedings of the Royal Society of London *452 (1996): 1207–1220.*

Simon, M. D.; L. O. Heflinger; and S. L. Ridgway. "Spin Stabilized Magnetic Levitation." American Journal of Physics *65 (1997): 286–292.*

*206. Levitating Mouse

A magnetic field of only a few tesla can lift and levitate nonmagnetic materials such as a drop of water or even a mouse. Graphite beads were first levitated in 1939. In 1991 began the parade of larger-object levitations.

All materials exhibit a magnetic response, however slight. Even a mouse has a nonzero magnetic susceptibility! Any electromagnetism text derives the pertinent expression, that the upward magnetic force acting on the magnetic material is $(\chi/\mu_0)V\nabla B^2$, where V is the material volume, χ is the magnetic susceptibility, μ_0 is the magnetic moment, and B is the magnetic field. The downward force is ρVg. Therefore $\nabla B^2 = 2\ \mu_0\ g(\rho/\chi)$, a condition easily obtainable in the lab.

Geim, A. "Everyone's Magnetism." Physics Today *51 (1998): 36–39.*

Chapter 7
Bodies in Motion

207. Superwoman

In the diagram, the woman (who weighs Mg) and the chair (which weighs mg) must be lifted by the woman pulling downward on the rope. Consider the ideal situation: an extensionless, massless rope, a massless pulley that does not hinder its free rotation, and a rigid support. Place a large imaginary box around the woman and the chair so that only the rope extends outside this box. (The imaginary box isolates the forces acting only inside this box from the external forces.) The rope goes around the pulley above and supports the box twice. By Newton's second law, in the vertical direction the downward pull of gravity (equal to the total weight, Mg + mg) must be exceeded by the total upward pull of the two rope segments to have a net force upward and an acceleration upward. The tension T upward in each supporting rope segment makes a total upward force $2T$, so $2T$ must be greater than Mg + mg for the system to accelerate upward. Therefore, for a 110-pound woman and a 10-pound chair, the woman must exert at least a 60-pound force on the rope, a feat easily accomplished.

208. Lifting Oneself by One's Bootstraps

In the diagram, the man (who weighs Mg) pulls upward on the rope with a force T, which produces a tension T in the rope. Consider the ideal situation: an extensionless, massless rope, a massless pulley that does not hinder its free rotation, and a rigid support above. Place an imaginary box around the man and the block so that only the rope extends outside to attach to the rigid support. (The imaginary box isolates the forces acting only inside this box from the external forces.) By Newton's second law, an upward acceleration begins when the upward tension T in the single support rope segment exceeds the downward pull of gravity on everything inside—that is, Mg + mg, the weight of the man and the box. Hence, when the man pulls with a force $T >$ Mg + mg, he and the box will rise together from the ground.

In an actual test, described by J. P. Drake in the *Scientific American* of October 20, 1917, a 190-pound man lifted not only himself this way but a 110-pound block as well.

Mott-Smith, M. Principles of Mechanics Simply Explained. *New York: Dover Publications, 1963, pp. 144–145.*

209. Springing into Action

The balance will read 100 pounds! When the object weighing 60 pounds was hung on the hook of the spring balance, the tension in the lower rope immediately decreased to 100 pounds minus 60 pounds, or 40 pounds. However, the sum of the downward forces exerted by the 60-pound object and the 40-pound tension in the rope still added up to 100 pounds. The 60-pound object took away some of the burden carried by the rope, but the total burden remained the same. Thus, if any object up to 100 pounds is hung on the hook, the reading will remain 100 pounds. If a 100-pound object is hung on it, the tension in the rope will become zero, the object having taken over the role of the rope. If more than 100 pounds are hung on the hook, the rope will become completely slack, and the reading will be equal to that of the weight of the object suspended from the hook.

210. The Monkey and the Bananas

The opposite external torques produced by the monkey and by the bananas about the axle of the pulley will cancel each other. The angular momentum L about the pulley axis is thus conserved as required by the law of conservation of angular momentum. Here L is initially zero, so it remains zero regardless of what the monkey does. Specifically, any upward movements of the monkey and the bananas must always be equal. Of course, if the monkey starts out lower than the bananas, the vertical distance between them will remain the same, resulting in one frustrated monkey. (We assume that the bananas never get high enough to wedge at the pulley!)

Looking into the details of the forces, you need to consider the tension along the rope, which must support the monkey's weight and provide the force for his acceleration up the rope on that side while supporting the bananas on the other side. Strictly speaking, an extensionless rope cannot increase its tension; however, you should assume that the extensionless rope has an identical tension at all points along the rope.

At the monkey's end, the last rope segment pulls upward on the monkey with the tension $T = (mg + ma)$—that is, with a force (mg) to support its weight plus the applied force equal to ma (the amount of tug of the monkey on the rope) to provide an acceleration upward. The same tension acts on the bananas at the other end of the rope to accelerate them upward equally. The monkey and the bananas will rise together.

211. Hourglass on a Balance

From the instant that the first grain of falling sand strikes the bottom of the hourglass to the instant the last grain of sand leaves the upper chamber, the force resulting from the impact of the falling stream remains constant and helps make the total weight equal to the weight of the hourglass before inversion. When the stream of sand begins to fall, the freely falling sand does not contribute to the weight, so there is slightly less weight registered for the first few hundredths of a second. As the last grains of falling sand strike, there is a short time interval when the weight exceeds the initial weight. For each grain of sand now striking the bottom, no longer is there a grain of sand leaving the upper chamber, so the hourglass weighs more.

Shen, K. Y., and B. L. Scott. "The Hourglass Problem." American Journal of Physics *53 (1985): 787.*

212. How Much Do I Weigh, Anyway?

The fluctuations result from the up-and-down movement of the center of gravity of the blood as the heart goes through its beat cycle. For a person weighing 165 pounds, the amplitude of the fluctuation is about 1 ounce. You can simulate this effect (with much greater results!) by standing on a bathroom scale and moving your arms up and down.

As you begin to step off the scale, you must slightly bend your knee or knees to take that first step. Most of your body is accelerating downward momentarily, so the scale no longer supports its full weight. Therefore the scale reading *decreases* slightly!

213. A Bumpmobile

The system has the same horizontal momentum just before and just after the collision of the hammer with the plank. Just before the moving hammer collides with the stationary plank, its momentum is in the direction of the plank. Just after the collision, the plank (with the woman aboard) is moving in the original direction of the hammer motion, and the hammer is now moving with the plank (ideally). The action has transferred the horizontal momentum from the hammer to the plank + woman + hammer to satisfy the law of conservation of momentum.

The friction with the floor plays two roles. First, the static friction prevents the plank from moving until the hammer strikes its blow. Second, the kinetic friction acting between the floor and the moving plank after the blow brings the moving system back to rest while transferring momentum to the earth.

Phillips, T. D. "Finding the External Force." American Journal of Physics *22 (1954): 583.*

214. The Wobbly Horse

At first the horse accelerates from rest but soon reaches an average speed, which remains constant until the horse nears the edge of the table. The initial acceleration from rest is in response to the horizontal net force applied through the thread by the weight hanging over the edge. The nearly constant average speed is the result of this constant horizontal applied force being matched by the static frictional force opposing the forward motion. When the wobbly horse approaches the edge so that the angle of the thread tends more toward the vertical, the normal force of the horse against the table increases significantly. This increases the static frictional force to bring the horse to rest just before the edge. What a smart horse!

Many people consider only the horizontal thread force and its decrease in value as the horse nears the edge. They fail to apply Newton's laws correctly, for even if this thread force becomes zero before the horse reaches the edge, the horse would still tumble over the edge! Newton's first law says that the horse should continue in its state of uniform motion in a straight line unless acted upon by a net external force. Here, the larger net external force is the static frictional one, which might have brought the horse to rest even if its value did not increase.

215. Two Cannons

The surprising answer is that no matter what the distance between the cannons and at what angle they are aimed, the shells will always collide in flight (neglecting air effects).

To understand why, turn off the force of gravity momentarily. The shells then travel along the straight-line path between the cannons and collide at the midpoint. Turn on the gravity and they fall equal distances to collide in midair once again.

216. The Law of Universal Gravitation

The formulation given is incomplete. Newton clearly stated that the inverse-square law for universal gravitation applies to mass particles rather than to extended bodies, with the distance d being the distance between two mass particles. Only in the case of radially symmetric spheres does d refer to the distance between their centers of mass—that is, their geometrical centers. In all other cases, one must integrate the force over the particle constituents of the extended bodies.

217. Balancing a Broom

No. The shorter part of the broom, the part containing the whisk, is heavier. The short part and the long handle balance because they exert equal and opposite torques about the point of support, not because they have equal weights. The center of gravity of the short part is closer to the point of support, so its weight (which may be assumed to be concentrated there) must be greater to provide the balancing torque. Think of two kids on a seesaw: for balance, the heavier kid must sit closer to the fulcrum.

218. *Vive la Différence*

The center of mass of a man is usually closer to the head than that of a woman. Therefore, a typical man cannot knock over the matchbox without moving his center of mass forward of the knees, thereby tipping over. In other words, the knees form the horizontal axis about which the center of mass provides a torque. As long as the torque rotates the person back toward his feet, the system does not tip over. Another way to state this condition relies on the center of mass being above the area of the support base defined by the toes and the knees.

Women also have an advantage over men when floating on their backs in water because their weight distribution tends to be significantly different. For men, the center of buoyancy is far separated from their center of gravity, the former in the chest region and the latter near the buttocks. For women, both centers are in the region of their abdomen. As a result, a man floats at a slight angle, with the upper torso farther out of the water than the lower torso. A woman tends to float level.

McFarland, E. "Center of Mass Revisited." Physics Teacher *21 (1983): 42.*

219. Balance Paradox

In both diagrams, links *AC* and *BD* are always vertical, and the bars *EF* and *GH*, which are rigidly mounted on the links, are always horizontal. Since *F* and *G* are at the same distance from the central axis, the objects on *EF* and *GH* move up and down through the same distance no matter where they are set on the bars.

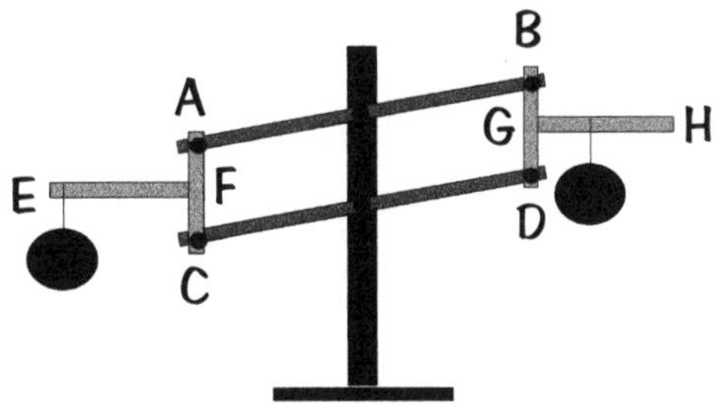

The weights of the objects being equal, the work done by gravity in lowering the object on *EF* must be equal to the work that could be extracted from the object on *GH* after it is raised. But the amount of work for rotational motion is the torque times the angular distance moved. Since both sides of the longer bars of the pantograph move through the same angular displacement, the opposite torques about the pivot pins must also be equal in magnitude. Therefore, the system remains in balance no matter where the two objects are placed along the horizontal bars on each side.

If we remove the bars *EF* and *GH* and attach trays to *A* and *B*, we obtain a balance with a very useful property: we do not have to be careful to place the object weighed or the weights at the center of the trays.

In fact, the parallelogram arrangement is the essential element in all balances whose pans are supported from below rather than by suspension from a beam above. The balance thus constructed is called a Roberval balance, after the French physicist and mathematician who invented it in 1669.

"A Balance." Little Stinkers section of Physics Teacher *3 (1965): 39.*

220. Tightrope Walker

The extra weight is of very little concern to the tightrope walker, who must prevent himself from falling off the wire. The heavy bar increases his moment of inertia about the tipping axis parallel to the wire, so that any tipping occurs much more slowly than without the bar. Hence there is a greater recovery time available to restore balance.

A physicist would place most of the mass of the bar near the ends, because the moment of inertia $I = mr^2$, where r is the distance from the rotation axis. A small mass out there is as effective as a much larger mass near the performer.

221. Balancing an Upright Stick

The statement that bodies with low centers of gravity are more stable than those with high centers of gravity applies only to situations involving static equilibrium. Under these conditions, a very small tilt from the upright position will move the vertical line of the center of gravity outside the contact area of the base, producing a net torque about a horizontal axis. Therefore, the taller stick falls over easily compared to the shorter pencil stub, which requires a larger tilt.

When balancing the stick on the tip of a finger, the finger can be moved to keep it underneath the center of gravity of the stick. The longer stick has a greater moment of inertia, so its angular rate of turning is smaller than for the shorter stick. You therefore have enough time to move your finger back under the center of gravity before the stick falls over.

222. Racing Rods

Contrary to most people's expectations, rod *A* reaches the lowest position before rod *B*. In fact, during the entire motion from highest to lowest position, rod *A* is always ahead of rod *B*.

There are many ways to analyze the behavior of the rods. For example, applying Newton's second law to torques, one derives that the angular acceleration is proportional to the ratio of the torque divided by the moment of inertia about the pivot point. The greater torque acting on rod *B* is not enough to compensate for its greater moment of inertia, so its angular acceleration is always smaller than the angular acceleration of rod *A*.

Hoffman, P. O. "A Mechanics Demonstration." American Journal of Physics *23 (1955): 624.*

223. Magic Fingers

You would expect the upper supporting finger to slip first because it seems to be supporting less weight. Its maximum static frictional force value would be less and therefore easier to exceed. However, by keeping the stick at the same angle, pushing inward equally at both sides momentarily increases the supporting force (and therefore the static friction) at the upper contact and allows the lower contact finger to move first.

224. The Soup Can Race

A liquid soup such as chicken noodle soup does not couple well (i.e., slips) with the inner wall of the can as it rolls down the incline. Therefore, most of its total kinetic energy of motion at each lower position along the incline will be translational kinetic energy, with very little as rotational kinetic energy. On the other hand, a more solid soup, such as cream of broccoli, will rotate along with the can, so that the rotational kinetic energy will be appreciable and less will be translational kinetic energy. Hence the more liquid soup will always have the greater translational velocity down the incline to win the race.

The mass and the radius of the can do not play a primary role in the rolling behavior of large-radius cans, but one must consider the proximity of the wall of the can to the fluid inside for viscosity coupling purposes. As the can radius becomes smaller, more and more of the liquid soup will try to roll with the same rotational motion as the can.

Stannard, C. R.; P. O. Thomas; and A. J. Telesca Jr. "A Ball with Pure Translational Motion?" Physics Teacher *30 (1992): 526.*

225. The Tippe Top

With respect to the person looking down on the top, the inverted tippe top and the upright tippe top spin in the same direction. However, since the top has turned over, its body rotation must have reversed itself! Considering rotations about the vertical axis only, the friction with the surface applied the necessary torque to accomplish this feat as the top turned over.

Barnes, G. "Tippe Top Thoughts." Physics Teacher *25 (1987): 200.*

Cohen, R. J. "The Tippe Top Revisited." American Journal of Physics *45 (1977): 12–17.*

226. The Mysterious "Rattleback" Stove

The misalignment of the long axis of the ellipsoid to the long axis of the flat top—that is, the body axis—contributes to the behavior. If the stone is spun in the "wrong" direction, the kinetic frictional force eventually brings the stone to rest rotationally, but the rocking rattle motion continues. When the downward rocking touches the table surface just right, the table exerts a small net rotational torque in the "right" direction, and the stone begins rotating. As long as the rocking motion continues, additional small net torques can ensue to continue turning the stone in the "right" direction against the opposing frictional force.

Walker, J. "The Mysterious 'Rattleback': A Stone That Spins in One Direction and Then Reverses." Scientific American *250 (1979): 172.*

227. The Case of the Mysterious Bullet

The bullets are identical except for the material of composition. Bullet *A* must have suffered an elastic collision with the target and rebounded, while bullet *B* embedded itself in the target. In the simplest case, the change in momentum of bullet *A* was twice the change in momentum for bullet *B* if the change in momentum for both bullets occurred during the same time interval. Then the force of impact by *A* would be twice the force by *B*.

228. Centers of Mass of a Triangle and a Cone

The center of mass of a right circular cone is located at a point *one-fourth* up the altitude of the cone. The reason for this lowering of the center of mass will become clear if we imagine the cone to be composed of thin triangular slices parallel to the largest triangular slice passing through the vertex. The center of mass of each such triangular slice lies one-third up the altitude of the slice. However, as the slices get smaller toward the outside of the cone, so do the altitudes and the centers of mass get closer and closer to the base of the cone. As a result, the center of mass of the whole cone is brought down to a point one-fourth the distance up its axis. The value one-fourth can be calculated using calculus.

229. Staying on Top

Several factors influence where the apples move during the shaking. No apple larger than the available space between lower apples can slip downward through that space, so if the lower apples do not move aside, the larger apples will remain above them. However, an apple smaller than the available space can easily drop downward. When this repositioning occurs many, many times in a bucket of apples of different sizes, then the larger apples will tend to be left behind at the upper echelons.

In the bigger picture, a system assumes its most stable position when its potential energy reaches a minimum. The center of gravity of the apples will be in its lowest position when the apples in the lower portion of the bucket become as densely packed as possible. The center of gravity will be the lowest if all the nooks and crannies in the lower portion are filled with small apples. As a result, the larger apples will tend to end up on top.

More surprisingly, even denser objects can be made to move upward in this fashion! The rising of rocks in the spring, although usually explained as due to the frost in the ground, is ultimately caused by perturbations that momentarily allow small grains to slip downward under the rock to inhibit a return to its initial position. Here they take the form of melting and freezing, but the same result could also come from shock and vibration. As another example of separation by size, where do we find the unpopped or partially popped corn kernels? At the bottom. Here, the higher density of the unpopped kernels is also a contributing factor.

Raybin, D. M. "The Stones of Spring and Summer." Physics Teacher *27 (1989): 500.*

230. Antigravity

Watch the marble carefully from the side and you'll see what really happens. As the marble is rolling toward the upper end, it actually descends slightly between the widening straws. The same effect can be observed when a double cone (i.e., a cone with two nappes) made from two plastic funnels is allowed to roll down a sloping double track cut from cardboard.

Edge, R. D. "String and Sticky Tape Experiments: An 'Antigravity' Experiment." Physics Teacher *16 (1978): 46.*

231. Which Path?

The ball rolling along ADC will reach the bottom first. True, the balls cover identical distances, and the accelerations along *AB* and *DC* as well as those along *AD* and *BC* are the same because of the same inclinations of the planes. However, the ball traveling along *DC* will have a high initial speed acquired during its rapid descent along *AD*. On the other hand, the ball traveling along the corresponding side *AB* will have a lower average speed, since its inital speed is zero.

232. Is the Shortest Path the Quickest Way Home?

The shortest time paths are shown. Comparing the three, the shortest time path is actually *PB*, not *PA* or *PC*. Most people think that the shortest time path will be the one with the horizontal tangent line at the bottom of the curve. Not so. If the coordinates of the final point are (p,q), the cycloid path will go through this point with an upward slope if $p/q > p/2$.

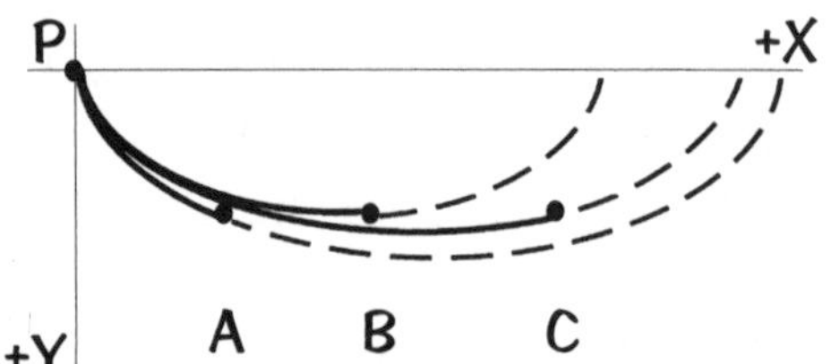

This path, *PB*, is the "brachistochrone" (i.e., shortest time), first solved by John Bernoulli (1667–1748), but the curve that provided the solution, the cycloid, was already known to Galileo as the archlike path traced out by a point on the rim of a rolling wheel. The cycloid is so lovely and led to so many controversies that it was dubbed the "Helen of Geometry."

The cycloid also has an amazing "tautochrone" (i.e., same time) property: A frictionless bead will reach the bottom in the same time interval no matter where on the curve the bead is released from rest!

Hoffman, D. T. "A Cycloid Race." Physics Teacher *29 (1991): 395.*

233. The Unrestrained Brachistochrone

Marble *B* wins the race by going down into the valleys and up over the hills. The horizontal component of the velocity of marble *B* at any moment is always equal to or greater than the horizontal velocity of marble *A*.

Remember that neither marble can leave its track, and neither marble slips.

Stork, D. G., and J. Yang. "The Unrestrained Brachistochrone." American Journal of Physics *54 (1986): 992.*

Zwicker, E. "High Road/Low Road." Physics Teacher *27 (1989): 293.*

*234. Tilting Rods

The bare rod will hit the floor first. The time of fall is determined by the angular acceleration, which is proportional to the ratio of the torque due to gravity and the moment of inertia. Both torques depend upon the distribution of the mass.

The angular acceleration of the bare uniform cross-section rod of mass m is $(3g/2L)\sin\alpha$, where L is the length of the rod and α is the angle between the rod and the wall. One learns immediately that just before the rod hits the floor (α equal to 90 degrees), the vertical downward acceleration at the end of the rod is $3/2\ g$—that is, greater than g!

For comparison, the angular acceleration of a rod with a point object of mass M attached at a distance d from the pivot is $a\,(1 + 2kq)/(1 + 3k\,q^2)$, where $k = M/m$, $q = d/L$, and $a = (3g/2L)\sin\alpha$, the angular acceleration of the bare rod. This result has several surprising consequences. For $q = 2/3$, thus placing the point mass two-thirds of the way from the pivot point, the angular acceleration will be the same as for the bare rod! For $q > 2/3$, the factor becomes less than 1, the time of fall becomes longer, and the rotational inertia effect increases more than the torque effect to produce our previous result. The opposite is true for $q < 2/3$.

Haber-Schaim, U. "On Qualitative Problems" (letter). Physics Teacher *30 (1992): 260.*

Hewitt, P. G. "Figuring Physics." Physics Teacher *30 (1992): 126.*

*235. Faster than Free Fall

The cup does accelerate greater than g, the acceleration of free fall. This result does not violate the expected free-fall behavior because the ruler is not in the state of free fall. In addition to the force of gravity acting on the ruler at the center of gravity, there is an upward force supporting the ruler at the point of contact with the floor. The total torque due to the gravity force plus the floor-contact force can produce a vertical component of acceleration downward greater than g at points along the ruler. Assuming that the ruler can be treated as a thin, uniform rod, the vertical component of the acceleration of the falling end is found to be $3/2\ g\sin^2\phi$, where ϕ is the angle the ruler makes with the vertical. This acceleration value increases as ϕ

increases and will exceed g when ϕ is about 35 degrees or less.

Try other variations: What happens when you reduce the starting angle or place the marble and the cup closer to the bottom end of the ruler? Will the ruler accelerate greater than g?

Edge, R. D. String and Sticky Tape Experiments. *College Park, Md.: American Association of Physics Teachers, 1987, experiment 1.50.*

Theron, W. F. D. "The 'Faster than Gravity' Demonstration Revisited." American Journal of Physics *56 (1988): 736.*

*236. Racing Cylinders

Let the cylinders roll without slipping down an inclined plane. At the bottom of the plane the total kinetic energies of the cylinders must be the same, since they descended through the same height—that is, the change in gravitational potential energy was the same for both. The total kinetic energy at the bottom (and all throughout the journey) consists of the translational part $1/2\ m\ v_{cm}^2$, corresponding to the motion of the center of mass, and the rotational part $1/2\ I\omega^2$, where I is the moment of inertia and $\omega = v_{cm}/R$ is the angular velocity of a cylinder of radius R. Setting the kinetic energies of the two cylinders equal, the greater moment of inertia of the hollow cylinder indicates a lower v_{cm} and vice versa. The hollow cylinder will roll slower all the way down.

*237. Friction Helping Motion

No, the conclusion is correct. For rolling without slipping, the point of contact with the ground—P, say—is instantaneously at rest. Then the cylinder rotates about a horizontal axis through P at any instant. We ignore the friction because its torque about this axis through P is zero—no lever arm. Using Newton's second law for torques about P, one determines that $2RF = (MR^2/2 + MR^2)\ (a/R)$, where R is the radius and a is the horizontal acceleration of the center of the cylinder. This equation yields $a = (4/3)\ F/M$. The static frictional force in the horizontal direction must be $F/3$ and in the same direction as the applied force! One must realize that the rolling cylinder pushes backward via static friction and so the ground responds according to Newton's third law.

Relland, R. J. "Two Fundamental Surprises." Physics Teacher *27 (1989): 326.*

Sherfinski, J. "Rotational Dynamics: Two Fundamental Issues." Physics Teacher *26 (1988): 290.*

*238. Obedient Spool

For all objects rolling without sliding, the point of contact with the floor—P, say—is instantaneously at rest. The horizontal axis of rotation will pass through P, so we must consider torques about this horizontal axis. There is no torque about this axis whenever the ribbon force line of action passes through the axis because there is then no lever arm. Pulling the ribbon at this critical angle (equal to r/R), the spool simply slides on the surface. When the force line angle exceeds this critical angle, the torque about the axis is clockwise and the spool rolls toward the observer. On the opposite side of the critical angle, the spool rolls away.

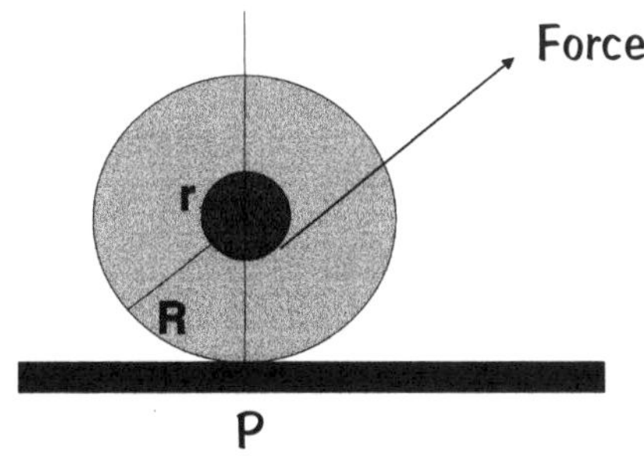

*239. "And the Winner Is . . ."

The solid cone beats the solid sphere! The determining factor is the numerical coefficient in the moment of inertia formula. The lower the coefficient value for the axis parallel to the rotation axis, the greater is the acceleration down the incline. The numerical coefficients for the moments of inertia are: (a) 1/2 for a solid cylinder about the cylinder axis; (b) 2/5 for a solid sphere about any diameter. For the right circular cone the corresponding value is 3/10, so the solid cone wins all races. To make the cone roll straight down the incline, fix an almost massless wire hoop having the base diameter near the apex end of the cone.

By the way, these three figures—cylinder, sphere, and cone, the latter two just fitting inside the first—are the three figures on Archimedes' gravestone. He was the first to determine their volume ratios: 3:2:1.

*240. Getting into the Swing of Things

The mechanism for starting a swing from rest is quite interesting. The child starts from an equilibrium position. At rest the point of support is directly above where the child grips the rope with her hands and her center of gravity. The tension in the rope T is vertical and is balanced by the child's weight W. The weight of the seat and rope is neglected.

Now we can see the real dilemma. The only external forces acting on the child are the tension in the rope and her weight, but they add up to zero. By Newton's first law, as long as the net external force is zero, no matter what the child does, her center of gravity is going to remain at rest. If she leans backward, her legs will go forward and up; if she leans forward, her legs will go backward. The center of gravity constantly readjusts its position in response to any changes in the body's configuration *relative to the body.*

We need a method to produce a net torque about the horizontal axis through the points of support with the frame. The tension in the rope will always point toward this axis, so the tension can never produce this torque. But the vertical-weight vector can have its line of action displaced from this rotation axis so that a lever arm exists. The child simply gives the rope a sudden backward jerk with just her arms, keeping the rest of her body rigid. By Newton's third law, her body moves forward slightly, and now the small clockwise torque about the support axis begins her motion backward.

To start from rest in the sitting position, the child leans back suddenly to acquire angular momentum about the center of mass, but with no external torques about the swing's pivot axis, the center of mass gets displaced from its initial position to start the swinging motion. For different approaches, see the references.

Curry, S. M. "How Children Swing." American Journal of Physics *44 (1976): 924.*

Gore, B. F. "The Child's Swing." American Journal of Physics *38 (1970): 378.*

*241. Pumping on a Swing in the Standing Position

A child standing on a swing can pump it in several different ways. What they have in common is that the child bends his knees at the end of the backward or the forward swing (or even at the end of both), and he straightens his knees in the middle of the backward, forward, or both swings, respectively. This knee-straightening and -bending motion has the effect of raising the child's center of gravity (*CG*) in the middle of the swing and lowering the *CG* at the end of the swing cycle.

To raise his *CG* in the middle of the swing, the child does work in two ways: (1) increasing his gravitational potential energy and (2) increasing his kinetic energy.

The kinetic energy is increased because the angular momentum about the horizontal axis of support does not change at the instant when the *CG* is raised. The torque due to a force whose line of action passes through the support axis is zero. The angular momentum is the product MVL, where M is the child's mass, V is his speed, and L is the distance of the *CG* from the axis. When L is made shorter, the speed V must increase so that the product MVL remains constant.

As the speed V becomes larger, so

does the child's kinetic energy 1/2 MV^2. This situation is similar to the action of a pirouetting skater as she pulls in her arms to increase her speed of rotation.

When the child lowers his *CG* at the end of the swing, his potential energy decreases. There is no kinetic energy change because he is then instantaneously at rest. His new *CG* position is now along the same arc as initially, but he is out at a greater angle, ready to begin the sequence anew. Over the swing cycle there has been a net gain in energy, which increased the amplitude of the swing, and this energy has been transferred from the child via his muscles.

Tea, P. L. Jr., and H. Falk. "Pumping on a Swing." American Journal of Physics *36 (1968): 1165.*

*242. Pumping on a Swing in the Sitting Position

Yes. In the standing position, bending and straightening the knees at the appropriate moments will pump the swing to greater amplitudes. In the sitting position, the motion of the center of gravity would hardly be affected. What the child actually does is very interesting. At the end of the backswing and the beginning of the forward swing the child swings her legs forward, thus rotating her body counterclockwise. Of course, her rotation cannot continue, for she would fall off the seat, so she stops it by pulling back on the rope.

Curry, S. M. "How Children Swing." American Journal of Physics *44 (1976): 924.*

*243. Spinning Wheel

No. Paradoxically, the student must push upward with his right hand and *downward* with his left!

First, we give the argument without using torques. Consider the state of motion of four mass elements in the tire. The one at the top has its velocity vector horizontal and straight away from the student, so a small velocity change to the left is required for the proposed rotation of the plane of the wheel. The one at the bottom has its velocity vector horizontal and straight toward his stomach, requiring a change to the right. The ones at the front and the back have vertical velocity vectors, down and up, which require no changes at all for the desired shift in the wheel's orientation.

The argument can be extended readily, by taking horizontal and vertical velocity components, to show that all mass elements in the upper half of the wheel need a velocity change to the left, while all those in the lower half need a velocity change to the right.

Since the only way to change the velocity of a mass element in a given direction is to apply a net force in that direction, the student must act through the axle, bearings, hub, and spokes by pushing upward with his right hand and downward with his left. Without using torques, we have discovered the peculiar "sideways" behavior of gyroscopic forces: To produce an effect in one plane, one must apply forces in the plane at right angles to the first plane.

Now the torque explanation. Initially, the spinning wheel has its horizontal angular momentum vector along the axis of the wheel, which we choose to align with the x axis. Pushing forward with the right hand and pulling backward with the left would decrease this x component of the angular momentum and increase the y component. But the applied torque is actually about the z axis! So the axle actually dips downward to increase the z component instead of following the desired course. To increase the y component of the angular momentum, one must apply a net torque about the y axis—that is, push upward with the right hand and downward with the left!

*244. Collision with a Massive Wall

The ball must rebound from the wall with a speed slightly less than the incident speed, even though the collision is elastic. Hence the elastic collision is only a good approximation to the ideal case of the object having the same kinetic energy before and after a collision. This effect is well known for gas molecules striking the ideal movable piston and thus doing work during the expansion of an ideal gas.

The momentum of the wall + earth, $P = MV$, is a definite quantity equal to $2mv$. The kinetic energy of the wall + earth can be expressed as $K = 1/2\ MV^2 = P^2/2M$. If P is not infinite, and M, which is the combined mass of the wall and the earth, becomes very large, the kinetic energy tends toward zero. This result shows that a massive object may have an appreciable momentum while at the same time possessing practically zero kinetic energy.

One may desire to estimate the amount of initial kinetic energy that has produced thermal and sound energies during the collision.

Macomber, H. K. "Massive Walls and the Conservation Laws: A Paradox." Physics Teacher *13 (1975): 28.*

*245. Executive Toy: Newton's Cradle

The toy illustrates the principles of conservation of momentum and energy. Suppose two right-hand balls are released from rest at a height h and strike the other balls with a velocity v. The total momentum just before the collision is therefore $2mv$. After the collision the three right-hand balls are stationary, and the two left-hand balls fly away with velocity v and total momentum $2mv$, exactly equal to the total momentum just before the collision.

Energy is also conserved: The total kinetic energy just after the collision is $1/2\, m\, v^2 + 1/2\, m\, v^2 = m\, v^2$, the same as the total kinetic energy just before the initial collision. Why didn't one ball pop out with velocity $2v$? The final momentum would be $2mv$. That's good. But the final kinetic energy would be $1/2\, m\, (2v)^2 = 2m\, v^2$.

There is a common conception that the principles of conservation of linear momentum and energy are enough to predict the behavior of the balls. But the two conservation laws provide only two equations for the final unknown velocities. Even if only three balls were used, two equations in three unknowns would result! To get a solution, we need some other guiding principle. One could choose to consider two-body collisions only, so that along the row, if the ball on the right with velocity v strikes a stationary neighbor, the initial ball comes to rest and the neighbor to the left starts moving with velocity v. The momentum gets transmitted along the row in this fashion. In this ideal example, all the energy and all the momentum are considered to be focused to the contact point between the adjacent spheres. In the real case, some energy transfers elsewhere in the sphere, the support, and into the surroundings.

More difficult problems can be worked: for example, an end ball of mass $2m$ strikes three aligned balls of mass m each. Just after the collision, the ball of mass $2m$ moves with velocity $v/3$, and the first ball, of mass m, moves at $4v/3$. Such calculations can be made for each collision along the row, and the problem will be solved completely.

Bose, S. K. "Remarks on a Well-Known Collision Experiment." American Journal of Physics *54 (1986): 660.*

Flansburg, L., and K. Hudnut. "Dynamic Solutions for Linear Elastic Collisions." American Journal of Physics *47 (1979): 911.*

Herrmann, F., and M. Seitz. "How Does the Ball-Chain Work?" American Journal of Physics *50 (1982): 977.*

*246. Hammering Away

In driving a stake into the ground we want to maximize the kinetic energy of the stake produced by the hammer blow. In forging a piece of metal we want to minimize the kinetic energy of the anvil and the hammer after the collision so the maximum amount of energy in the hammer blow becomes available to deform the piece of metal. The two cases are opposite situations.

Consider the completely inelastic collision first, the one without any rebound. This type of collision is best for the forging operation because the hammer would come to rest on the anvil with the metal piece.

The initial kinetic energy of the moving hammer is 1/2 $M_1 v_1$, and the total kinetic energy of the two colliding objects immediately after the collision is 1/2 $(M_1 + M_2) v^2$, where M_2 is the mass of the anvil (plus the metal piece). By taking the difference in kinetic energy after and before the collision and applying the conservation of momentum, one derives that the fraction of the initial kinetic energy now available for deforming the piece of metal is proportional to the ratio $M_2/(M_1 + M_2)$. Clearly, to make the fraction close to 1, the anvil mass M_2 should be much more massive than the hammer. In reality, most of this available energy becomes heat energy during the collision!

In driving the stake we want to maximize the kinetic energy of the stake during the collision, which means we must try to maximize the hammer mass M_1 with respect to the stake mass M_2. The ratio given above is only valid for the collision in which the hammer and the stake move together—that is, the completely inelastic collision—but it does give the right prediction here: make the mass of the hammer as large as possible.

To do even better, consider the perfectly elastic collision. Then all the initial kinetic energy of the hammer is transferred to the moving stake, and the friction of the ground must do more work to bring the faster-moving stake to rest. The stake penetrates deeper. In the general case, the kinetic energy transferred to the stake depends upon the mass ratio given above times e^2, where e is the coefficient of restitution ($e = 1$ for elastic, and $e = 0$ for completely inelastic collision).

Hartog, J. P. D. Mechanics. *New York: Dover Publications, 1961, pp. 291–292.*

Miller, J. S. "Observations on a Pile Driver." American Journal of Physics *22 (1954): 409.*

*247. Velocity Amplification

Assume that the mass of the small ball is negligible compared to the mass of the large ball, and that the collisions are elastic—ball to ball and ball to floor. One can approximate these ideal conditions by using superballs of two different sizes. During the elastic collision with the floor, the bottom ball—the large one—turns around and begins moving upward and would leave the floor with the velocity value v, the same speed it had just before its downward collision. Simultaneously, the top ball is about to collide with the bottom ball. From the point of view of the big ball, the small one is approaching the large one with a speed $2v$ (the small ball's speed plus the large ball's speed). When the small ball hits the other, it rebounds with velocity $2v$ upward with respect to this large ball. But let's not forget that the big ball is actually moving upward with a speed v, so the small ball is initially moving upward with the speed $3v$ with respect to the floor just after the collision. The height attained goes as the square of the initial speed component upward: so a three-times-faster ball goes nine times as high as its starting height! In more realistic cases, wherein the collisions are never elastic, the small ball can rebound at least four times its original height above the floor.

With a three-or-more-ball configuration, the effects are even more dramatic. In dropping a vertical stack of three balls with the smallest one on top, the small ball could theoretically approach a maximum height of forty-nine times the release height. The ideal height ratio (the final height achieved divided by the initial drop height) turns out to be $(2^N - 1)^2$, where N is the number of balls. This ratio increases so rapidly with N that one could put a ball on the moon by dropping a stack of fifteen balls from a height of only one meter! If only nature cooperated and followed the ideal case exactly!

Carpenter D. R. Jr.; D. J. Rehbein; and J. J. Barometti. "Ban™ Deodorant Ball Mortar." Physics Teacher *26 (1988): 522.*

Harter, W. G., Class of. "Velocity Amplification in Collision Experiments including Superballs." American Journal of Physics *39 (1971): 656.*

Spradley, J. L. "Velocity Amplification in Vertical Collisions." American Journal of Physics *55 (1987): 183.*

Stroink, G. "Superball Problem." Physics Teacher *21 (1983): 466.*

*248. Superball Bounce

Many physicists will respond, "It had to start with backspin," which is incorrect. First consider a simpler example. Let the superball approach the floor with a forward velocity and no spin. After bouncing from the floor the superball is still traveling forward, but more slowly. What is the direction of the spin? The superball will have topspin, with the top of the ball turning toward the forward direction.

The original problem is the time-reversed situation of this easy example. Newton's laws obey time-reversal invariance, but we need to ignore the heat production and the internal vibrations of the ball, etc. Then we see that the ball for the original problem must possess initial topspin to reach a final state of no spin and forward velocity after the bounce.

The detailed solution requires the application of the conservation of energy and of linear and angular momentum. The solution here for the initial problem demands that the momentum direction of the superball during the collision with the floor not produce a topspin torque—that is, its momentum vector direction be on the appropriate side of the axis of rotation during contact, a condition guaranteed by the dynamics of the ultraelastic collision.

One can then progress to showing that after the second bounce the superball is now in the same state as the initial state!

Bridges, R. "The Spin of a Bouncing Superball." Physics Education *26 (1991): 350–354.*

Crawford, F. S. "Superball and Time-Reversal Invariance." American Journal of Physics *50 (1982): 856.*

Garwin, R. L. "Kinematics of the Ultraelastic Rough Ball." American Journal of Physics *37 (1969): 88.*

*249. Ring Pendulum

As long as the cuts of the hoop are symmetrical, the periods of the remaining objects are all the same! An additional surprise is that even when practically all the hoop is cut off symmetrically, so that only a tiny piece remains, the period still does not change. One needs to show mathematically that the restoring torque and the moment of inertia have the same dependence on the distance from the rotation axis, so that a change in one quantity equals the change in the other.

*250. One Strange Pendulum

The amplitude of the simple pendulum swing will increase rapidly because this system is behaving as a parametric oscillator. Parametric resonance is strongest if the frequency of the parameter varying with time—that is, the support location—is twice the natural frequency of the driven system. That is, the energy transfer is best at $2f_0$ and is significantly less at other frequencies $2f_0/n$, where n is an integer.

For the simple pendulum with symmetric behavior on either side of the vertical line through the support point, one can appreciate that any effective energy-transfer scheme that occurs on the left half of the motion should also occur on the right half. Indeed, for this parametric example and for the more familiar example of a child on a swing, pumping in energy at $2f_0$ is quite efficient. Two parents at opposite ends of the swing arc can certainly increase the amplitude to the child's delight!

Landau, L. D., and E. M. Lifshitz. Mechanics. *Reading, Mass.: Addison-Wesley, 1969, pp. 80–84.*

Chapter 8 Stairway to Heaven

251. The I Beam

Take a wooden beam and support each end. If you now hang a few objects along the length of the beam, the beam will bend to support the loading. The top layers of the beam will be compressed to a slightly shorter length, while the bottom layers will be lengthened by the tension force. In between there is a neutral wood layer that will remain the same length and is useful solely in connecting the top and the bottom together.

Steel is more expensive than wood, and much denser. When girders are made from steel, most of the material must be placed where it does the most good—that is, there should be very little steel in the middle, near the neutral layer.

252. The Aluminum Tube

The solid rod is much harder to bend because there is more material to stretch at the bend location. In better words, the rod requires more energy to produce the same bend as the tube because more atoms are involved.

253. Two Pulleys

When the pulleys rotate clockwise, the belt will wrap itself around a greater portion of the pulleys' perimeters, increasing the coupling between them and therefore the power transmitted.

254. The Tensegrity Structure

The compression in the rods is produced by the tension in the wires. The wires cannot take a compression but are very strong under extension. Overall, the forces acting in all directions at any point sum to zero net force because the stable structure is not accelerating (via Newton's second law).

Large tensegrity structures can be seen at some art museums in their outdoor art gardens, where tensegrity towers more than 20 meters tall have been erected. You can build small tensegrity structures of less than one meter in length by starting with a suitable cardboard box with holes cut through at appropriate locations to temporarily help support the rods and wire. You then cut away the box to leave the finished product.

255. Vertical Crush

Like all materials, the bricks are elastic to some extent for a small amount of compression. The mortar between the rows of bricks providing the transfer of the load from above to the row of bricks below is compressible also. However, there is a tendency for the mortar on one side of the building to compress more than on the opposite side, so stability problems arise and continue to grow with time.

There are other obvious effects to be seen by the educated eye. For example, upon close inspection of many buildings, you can also discern more cracks on the sunny side of the building than on the shady side because the building expands and contracts differently as the temperature changes occur. These cracked-brick regions tend to be weaker.

Gordon, J. E. Structures, or Why Things Don't Fall Down. *New York: Da Capo Press, 1978, pp. 172–173.*

256. The Boat on High!

No. The weight of the water displaced (Archimedes' principle) equals the weight of the boat, so this displaced water has moved upcanal and downcanal. The engineer only needs to account for the weight of the water on the bridge when no boats are present.

257. Double the Trouble?

Even though the material of the strings is the same and therefore their strength/length ratio is the same value, the longer string requires about twice the energy to be broken than the shorter string. Why? Because twice the number of atomic bonds must be "stretched." Therefore, if either string breaks, the shorter one will break first. The longer string is termed more resilient.

By the way, you might want to consider the question of how a fisherman can catch a 50-pound fish with a 10-pound test fishing line!

Gordon, J. E. Structures, or Why Things Don't Fall Down. *New York: Da Capo Press, 1978, pp. 89–90, 139–140.*

258. Boat's Anchor

The chain has a maximum force along its length that can be sustained without breaking. If the chain is jerked suddenly, its own mass is added to the mass of the anchor, and the chance of exceeding this maximum force becomes higher. Therefore, well-trained sailors will pull gently on the anchor chain to hoist up the anchor.

259. Two Bolts

The bolt heads will remain at the same distance, and it doesn't matter which bolt is held stationary.

As long as the threads are meshed, a clockwise movement of bolt *B* around *A*, viewed from the bolt head end, is the same as a counterclockwise movement of *A* around *B*. While *B* is moving up the threads of *A* toward the head of *A*, bolt *A* is moving down the threads of *B* away from the head of *B*. The movement of the two threads cancels out. (If you don't have two identical bolts at hand, you can get the feel of the problem using one finger of each hand.)

260. Tree Branching

A few calculation estimates will demonstrate why branching patterns are so widespread in the living world. Set the distance between adjacent dots in the given diagrams equal to one distance unit. Then the total length for all paths, each starting at the center and terminating at a particular dot (a leaf), is 90 units in branching pattern (a) and 233.1 units in explosive pattern (b). But the average length of a path is 3.67 units in (a) and 3.37 units in (b). So even though the average path length is a little longer in (a), the branching pattern in (a) has a much shorter total length than the explosive pattern in (b). When energy considerations are accounted for, the branching pattern will win out easily. Thus, tree branches, blood vessels, rivers, and even subway routes are all examples of branching patterns.

261. Hurricane Winds

No. The 120 mile-per-hour wind has four times the force of the 60-mile-per-hour wind. The force of the wind varies as the square of the wind speed because both the mass of air hitting the house is more and the speed at which it hits is more. Doubling the wind speed doubles the mass per second at twice the speed, or four times as much momentum per second hitting the house.

However, for winds of low speed, the analysis can be quite different, depending upon whether there is laminar flow around the building or whether there are pockets of turbulence. In some cases the air flow must be considered as a whole entity on the macroscale instead of using the molecular view to achieve agreement with empirical data.

Epstein, L. "Wind Force and Wind Speed." Physics Teacher *29 (1991): 196–197.*

262. The Structural Engineer

The structural engineer was correct. Houses are built for stiffness, not for strength, because this goal is cheaper and easier. The structural shape is to remain fixed—that is, stiff. Houses do not carry significant loads, so the wall and floor strengths are not major concerns if they meet the building code standards until a special circumstance demands attention such as installing a water bed or a heavy load in the middle of a room on the second floor. If all the people at the party stand in the middle of a large upper-floor area and begin jumping, well, then . . .

263. My Arteries Are Stiff!

Take a close look at the graph showing the response of several materials to an applied force. Do you see how increasing the applied force to rubber or collagen or elastin, the latter two being fibers in the arterial wall, results in stretching the material? Now examine the stretch of the artery tissue. Notice how increasing the applied force results in very little stretch at first, then a sudden large stretch occurs as the applied force increases just beyond a threshold.

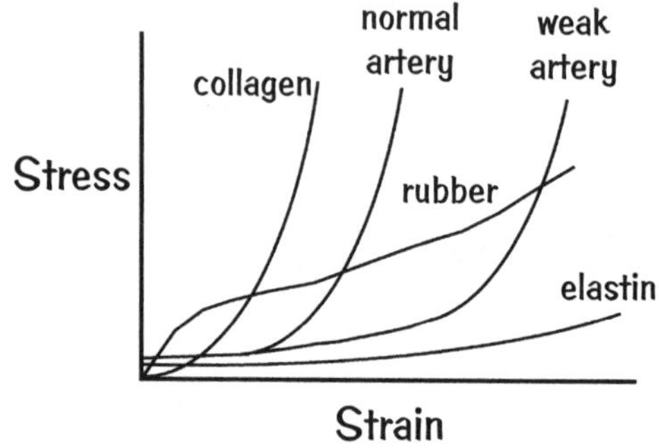

If the arterial walls were less stiff than they are, they would bulge outward whenever the blood pressure rises during the heartbeat. Such medical cases—called aneurysms—are abnormal, and could lead to the rupture of the artery. In the healthy artery, there is just enough "give" in the arterial wall to smooth out some of the pressure fluctuations.

Gordon, J. E. Structures, or Why Things Don't Fall Down. *New York: Da Capo Press, 1978, pp. 155–162.*

264. The Archery Bow

With the normal shooting of an arrow from the bow, most of the energy in the bow-arrow system is delivered to the initial flight of the arrow (i.e., the acceleration until the arrow leaves contact with the string). Very little residual energy must be dissipated by the bow. Without the arrow in place, all the energy must be released by the bow, and self-destruction can occur!

265. The Pork Sausage Mystery

Along the length. The skin of the sausage, like the skin of any pressure vessel, not only contains the fluids and other materials inside but also "carries the stress," the force per unit area. For cylindrical shapes of uniform material thickness, the circumferential stress capability before rupture is nearly twice the longitudinal value. So the rupture usually occurs along the length. Other types of tubes behave similarly, such as plastic tubing, metal piping, blood vessels (e.g., the aorta), and rifle barrels.

266. My Car Is a Steel Box!

The old chassis approach produced cars that were not stiff enough to limit torsional effects. That is, different parts of the chassis and body would bend different amounts, a result called differential bending, making the suspension requirements vary enormously under the different conditions of normal driving.

The modern body shell—the steel box—is essentially a large torsion box that is both strong and very stiff. In fact, for a torsion box structure the twisting resistance increases as the *square of the area* of cross section. Consequently, the torsion response is less erratic, and better suspension becomes possible because the ranges of physical operating parameters are limited better.

267. Balloon Structure

The air pressure inside the bubble must be kept slightly above atmospheric pressure to support the skin of the bubble over the stadium or the tennis court. Instead of a rigid support, like most structures possess, the compressible air inside does this job. Suppose that the balloon skin has a weight of 1 pound per 10 square feet. For equilibrium, the upward force of 1 pound acting against every 10 square feet is needed to support the roof. Converted to pounds per square inch, this pressure is only 0.007 lb/in^2, an incredibly small increase in the ambient air pressure inside the balloon roof! Just a few small fans can handle this. Of course, the doors must be kept closed.

268. The Open Truss

The open 3-D truss is composed of open tetrahedra, whose sides are open triangles that try to maintain their shape. Any other fundamental 3-D open trusses, such as a "rectangular box" frame, can "hinge" to become smaller in volume because they are not composed of triangles for all their sides. The usual examples of the 2-D triangle and the 2-D rectangle with pins at their corners can be used to demonstrate the stiffness of the open triangular shape versus the "give" of the open rectangular shape.

In addition, some of the edges of the open truss triangles will be in compression, and others will be in extension. The appropriate materials for these edges can be selected to optimize the strength-to-weight ratio for the conditions of the given application.

*269. Jumping Fleas

A human cannot jump many times his or her own height because human leg muscle strength does not allow us to generate a large enough initial velocity upward. The height of the jump $h = F\,s\,/\,m\,g$, where F is the average force against the ground needed by a creature beginning from a crouched position to raise the center of gravity a distance s before leaving the ground. Here m is the mass of the organism and g is the acceleration of gravity at the surface of the earth. In better words, the initial work done ($F\,s$) to produce an initial upward velocity is converted into the potential energy change ($m\,g\,h$).

Now we can understand why small animals can jump so high. Assume that the animal's strength F is proportional to the cross-sectional area of its muscles. Then F is proportional to L^2, where L is the animal's linear size. Therefore, the acceleration F/m is proportional to $L^2/L^3 = 1/L$. Since s is proportional to L, the height h is independent of the animal's size. Thus a flea blown up to human sizes could still jump only a few feet off the ground.

Or could it? The giant flea would collapse under its own weight, which would be about a thousand times greater, while its muscle and skeleton cross section would be only a hundred times greater. Apparently science fiction writers forget this problem when they try to scare us with giant insects.

*270. The Scaling of Animals

Doubling the bone diameter only makes the bone four times as strong (on the average), but making the diameter $2\sqrt{2}$ times larger than the original bone diameter will support the weight. The ribs need the $2\sqrt{2}$ factor because they are subjected to bending loads also.

Remarkably, the vertebrae only need to be twice as large in diameter because they are mostly under compression and press against each other (with a compression disk in between). Older bone has a greater crushing strength but becomes more brittle when subjected to bending and twisting forces.

*271. A Staircase to Infinity

The standard solution: It can. And not only more, but as much as we please! Of course, one must not invoke the crushing strength limit in either of the solutions below!

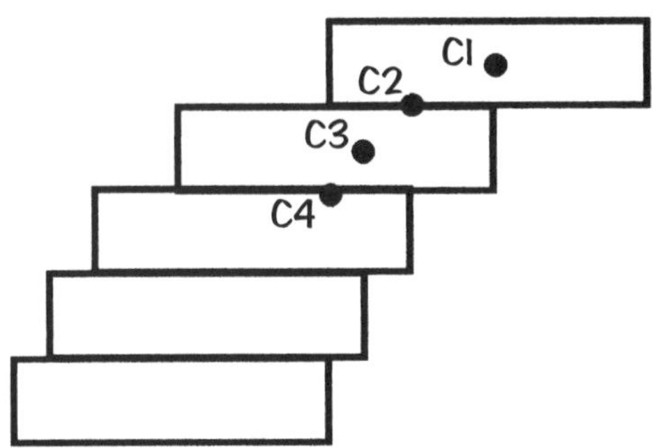

In the diagram, if one brick is placed on another, the upper brick will not fall if its center of gravity is anywhere above the next lower brick. The greatest offset, equal to half a brick's length, is obtained when the center of gravity (c.o.g.) of the upper brick C_1 lies directly above the end of the brick below.

Let's see how the structure is analyzed. We start with a few bricks and insert an additional new bottom brick each time. How can two bricks be placed on a third one to obtain maximum offset? The combined c.o.g. of the two top bricks is at C_2, one-fourth of the brick's length from the end of the second brick. Position C_2 over the end of the bottom brick. The second brick is offset one-fourth of a brick's length over the bottom brick.

To place these three bricks with maximum offset on a fourth one, place the c.o.g. of the three bricks, C_3, at the end of the new bottom brick. How is the c.o.g. determined? By equating the clockwise torque of the top two bricks about the axis through C_3 to the counterclockwise torque of the third brick. One then carries forward this series of steps ad infinitum. One finally arrives at the total offset of the top brick over the bottom brick as an infinite series: Total offset = $L/2$ (1 + 1/2 + 1/3 + 1/4 + 1/5 + . . .), where L is the length of one brick. The sum in parentheses is the famous harmonic series that does not converge to any noninfinite number—that is, the sum will be greater than any finite number.

An alternative solution: Stack the bricks on the lower bricks so the stacked bricks extend outward in *both* directions from the center to balance the torques. There will be a vertical line of symmetry as each brick to the left is balanced by an identically placed brick to the right. Both sides can extend to infinity.

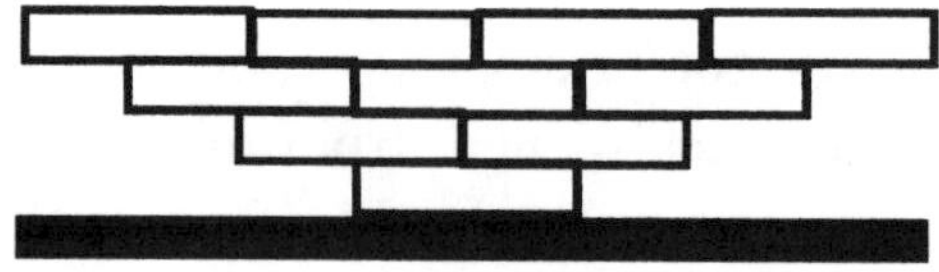

*272. Cowboy Lasso

The rotation of the nearly circular rope loop in a vertical plane is driven by a hand moving a section of the rope that is not part of the loop in a circular motion that starts out almost perpendicular to the lasso for a small loop. As the loop gets bigger, the length of the rope section held by the hand increases to the loop, and the angle between the rope section and the loop becomes much less perpendicular, rapidly approaching the vertical plane of the larger loop.

All around the vertical circular loop, the net force on a small rope segment must be directed radially inward in this ideal case to produce the inward radial acceleration for movement around the circle. This net radial force inward is the vector sum of three force types: the tension forces by the two adjacent rope segments, one on each side of the given rope segment, the radial component of the gravitational force, and the radial component of the bending force due to the rope stiffness or resistance to lateral deformation. For a circular loop, any change in the gravitational force radial component with angular position around the loop must be balanced by changes in the tension forces if the radial component of the bending force is assumed to be a constant.

The gravitational force radial component varies as cos α, where α is the angle measured from the downward vertical. Therefore, the tension force along the rope must also vary as cos α, with the maximum tension at the bottom of the loop. In addition, this tension force will increase with the rotation speed of the loop. Is there a minimum rotational speed that will keep the rope in a circle? Yes. When $v = \sqrt{Ra}$, the correct radially inward acceleration a is produced by the three forces listed above, with v being the tangential velocity and R being the radius of the loop. If v falls below this minimum value, then the circle collapses.

Chapter 9 Life in the Fast Lane

273. The Baby Carriage

Yes. Over a given distance the 1-foot wheel rotates twice as many times as the larger wheel, the result being more work against the friction in the bearings at the axle.

Another consideration would be the pebbles in the road or path. The horizontal force needed to push the 1-foot wheel over the pebble is greater than the force needed for the 2-foot wheel. One could draw the force diagram and examine the vertical and horizontal force components for moving the wheel over the pebble. This effect is one reason why the famous Conestoga wagons, used in the early westward expansion, had such large wheels. Also, fewer turns meant less wear on the axles.

274. The Falling Cyclist

By steering into the direction of the fall, the cyclist follows a curved path of such a radius as to generate enough centrifugal force to bring herself and the bicycle upright again. Once in the vertical position, the cyclist turns the handlebars over to get closer to the original direction. Before this maneuver, the cyclist and the rest of the bicycle had usually swung into line behind the front wheel because of a castering effect.

There is a tendency for the cyclist to oversteer. In either case, to get out of the initial curve the cyclist is forced to go into another curve on the other side of the original direction. Thus she progresses by a series of arcs that at high speed become almost imperceptible. After all, at high speed the cyclist can afford the luxury of a large radius of curvature r—an almost straight path—since she gets a sufficient centrifugal force from a large v^2 term in the numerator of $F_{centrif} = m\,v^2/r$.

Kirshner, D. "Some Nonexplanations of Bicycle Stability." American Journal of Physics *48 (1980): 36–38.*

275. Sudden Stops

The frictional force acting between two bodies is directly proportional to the force pushing them together and to the frictional coefficient of one in contact with the other. When the brakes are applied, the car pitches forward because the four wheels are slowing their forward motion as the car body continues onward. The connections between the wheels and the body finally act on the body to slow it down also. The center of mass of the car, being above the center height of the front wheels, exerts a rotational torque on the car trying to flip the car over the front wheels. Fortunately, most (but not all) of this dangerous torque is counteracted by the torque of the gravity acting at the center of mass also.

The net torque about the front-wheel axis tips the front end of the car down noticeably during a hard stop. Effectively, during this hard stop, an additional force of nearly 10 percent of the weight of the car may be pushing down on the front wheels, while the rear wheels will support less than before. The front-wheel brakes will need to exert most of the braking force (as much as 65%) so that the tires can "tell the road" via Newton's third law to push against them hard enough.

Hafner, E. "Why Does an Accelerating Car Tilt Upward?" Physics Teacher *16 (1978): 122.*

Whitmore, D. P., and T. J. Alleman. "Effect of Weight Transfer on a Vehicle's Stopping Distance." American Journal of Physics *47 (1979): 89–92.*

276. Braking

The two cases are very different. On the level road, there is no jolt as the brake is applied at zero velocity. On the uphill slope, a heavy jolt is felt as the brake is applied at zero velocity, because there is an abrupt change in the acceleration.

277. Car Surprise

The white car (front wheels locked) will descend with its front forward, and the black car will turn around so its locked rear wheels are forward!

The friction between the rolling wheels and the surface is static friction (each rolling wheel is momentarily at rest where it touches the surface). The locked wheels will be sliding, so these wheels experience sliding friction, which will be smaller for contact between the same two materials. The black car will start out with its front end forward but will not slide true. As the car turns a little, the net torque produced by the static friction at the front and the sliding friction at the rear will continue to turn the car, so the back end will have the opportunity to swing around to the front.

Unruh, W. G. "Instability in Automobile Braking." American Journal of Physics *52 (1984): 903–909.*

278. Engine Brakes

The braking action is greatest in low gear because the engine turns fastest (at any given speed) when in low gear. Kinetic energy is converted into the thermal energy of engine friction faster when the engine is turning faster. Engine braking to slow the speed of descent is a good alternative to normal braking to convert kinetic energy into thermal energy at the brakes.

279. The Transmission

The internal-combustion engine develops very little torque (twisting force) at low speeds, so this type of engine will stall very easily at rotational speeds below about 300 revolutions per minute. Its torque is so small at low rpm that any appreciable load will stall the engine. Therefore, a clutch is needed to disengage the engine from the transmission gears and to gradually engage the load until the rotation speed reaches more than 1,000 rpm or so, at which point useful torque is developed. A steam engine and an electric car can develop nearly full torque from a standstill!

280. The Groovy Tire

The tread on tires slightly reduces their grip on the road under dry conditions—less rubber is in contact with the road. Normally, the static frictional force is independent of the area of contact for rigid solids in contact with each other, but tires are not rigid. The evidence shows that a smooth tire will stop a car on dry pavement in less distance than tires with a good tread!

The grooved treads are designed for wet roads so that the water can move into the crevices and the rubber can contact the road without a thin film of water in between. On the whole, the sacrifice of a little grip on dry roads is paid back by the better behavior on the wet roads.

Brake linings are smooth and not grooved in order to maximize the contact area because the material is not a rigid solid but a solid that "gives" more at higher temperatures. The "slicks" on drag racers are "burned in" before the start to increase the "stickiness" of the tire contact with the raceway—that is, increase the static friction coefficient and the maximal value of the static friction before slipping.

Logue, L. J. "Automobile Stopping Distances." Physics Teacher *17 (1979): 318–320.*

Smith, R. C. "General Physics and the Automobile Tire." American Journal of Physics *46 (1978): 858–859.*

281. The Strong Wind

The moment the wheels start sliding instead of rolling, the static frictional force between the wheels and the road changes into the sliding frictional force, which is a smaller force for the same two materials in contact. The sideward wind force from the left can now exceed the maximum sliding frictional force of the tires against the road and accelerate the car into the lane to the right.

282. Wheels

The tangentially mounted spokes of bicycles carry two kinds of load: radial, by supporting the hub, which in turn supports the frame and the cyclist; and tangential, by resisting the twisting forces transmitted to the sprocket wheel by the chain (usually the rear wheel) and to the tires by the brakes (either or both wheels). To be able to carry tangential loads in either direction, the spokes are tangential to the hub in both forward and backward directions.

Wheels with radial spokes that carry mainly radial loads did not appear until about 2000 B.C.E. on chariots in Syria and Egypt. They became universal on carriages and wagons, where the source of locomotion was outside the vehicle.

Krasner, S. "Why Wheels Work: A Second Version." Physics Teacher *30 (1992): 212–215.*

283. Newton's Paradox

A correct application of Newton's second law of motion resolves this paradox. First consider an imaginary box around the wagon. Then ask what horizontal force or forces act from outside this imaginary box. If the sum of these external forces in the horizontal direction is nonzero, then there will be an acceleration in the direction of this net force. In this problem the rope is pulling forward on the wagon and provides the net force forward.

Confusion often occurs when trying to apply Newton's laws without properly identifying only the external forces acting on the object of concern. For example, if one isolates the horse, one learns immediately that the actual external force accelerating this noble steed forward is the static frictional force of the road upon the hooves, a force that can be greater than the backward pull of the rope by the wagon.

284. The Obedient Wagons

Each wagon's wheels take the same path as the wheels of the wagon ahead. In other words, no wagon decides to take a shortcut, but maintains its position on a circular arc. Since the wagons are identical, the tow bar angles are all the same, leading to a circular arc.

This problem was posed to one of us (F. P.) by Richard Feynman in his car on the way to Malibu, California, from CalTech in 1967, after he had watched connected luggage carts follow one another to the plane on the previous day at the airport.

285. The Escalator

As more people board the ascending escalator, the speed should slow down as the motors maintain a constant power level—that is, work rate. The real systems, however, adjust their work output in an attempt to maintain a nearly constant speed.

286. Roller Coasters

Each roller-coaster rider receives a different ride. On going over a hill, for example, the coaster does not gain any speed until after its center of mass reaches the peak. The front riders are already on a slow descent, so they experience a delayed acceleration; the middle riders are near the peak and are beginning to accelerate down; and the rear riders experience an increase in speed on the way up the hill. The reverse behavior occurs in the valleys.

287. Clothoid Loop

The roller coaster slows down when moving up the loop as gravitational potential energy grows at the expense of kinetic energy. The clothoid loop has two advantages over the circular loop. The sharper turn at the top results in a greater radial acceleration to keep riders in the cars even though they are moving slowly. And the slower speeds required to make the loop reduce the tremendous acceleration normally experienced at the end of a circular loop to a more manageable value.

288. Turning the Corner

When an automobile turns a corner, all the wheels slip a little. The outermost wheel on the turn always travels a longer distance around the curve, which has the greater radius. The front wheels on practically all cars (and also the rear wheels on all-wheel-drive vehicles) are designed to minimize the slipping by allowing the two wheels to point in slightly different directions, as needed, and by letting the outer wheel rotate faster. Some slipping still occurs at each front wheel because the optimum condition can be matched for only a very small region of contact of the tire with the road. The other sections of the tire in contact will experience some slippage. The situation for the rear wheels is not as good, even though they can rotate at different speeds, because they remain parallel to each other.

289. The Mighty Automobile

An automobile traveling at 50 miles per hour (about 75 kilometers per hour) will experience wind resistance that can be matched by a 20-horsepower (14.8-kilowatt) engine to keep the car velocity constant. But an acceleration with this small engine on a massive car will be quite sluggish, and perhaps dangerous, when the driver needs to accelerate from a stop or needs to pass another vehicle. Therefore, at least a 60-horsepower engine is recommended, but 200 horsepower and greater just adds a lot of vroom!

290. Front-Wheel-Drive Cars

Any vehicle with a lot of weight above the drive wheels will have better traction in the snow. The larger normal force will allow the maximum static frictional force value to be greater, so the wheel can push horizontally against the snow with a greater force before slipping occurs. Pickup trucks with rear-wheel drive can be loaded with sandbags to increase their weight above the drive wheels in order to improve the traction in snow or mud.

291. The Hikers

Putting the denser items higher up in the backpack is good physics. The higher the center of gravity of the backpack, the smaller the forward bending angle at the waist of the torso that the hiker needs to put his center of gravity above his feet. The smaller bending angle means less strain on the stomach and back muscles. Certain native tribespeople have perfected the feat of carrying heavy loads directly on their heads so that no forward bend is required.

292. The Fastest Animal Runners

The cheetah and the pronghorn deer certainly weigh much less than the elephant, so their legs are considerably less massive. In addition, their bodies flex quite readily, so they can extend their legs forward and backward more than most animals. Because the leg-muscle strength increases with the cross-sectional area while the mass increases with the volume, heavier animals tend to lose out in leg strength per kilogram. Thus the elephant has heavier legs and bigger leg muscles, but the leg-muscle strength per kilogram of leg mass is significantly below the values for the cheetah and the pronghorn deer. So even without all the mass above the legs, the elephant would lose the race.

293. The Oscillating Board

The friction between the rotating shafts and the board is sufficient to drive the board periodically left and right. As shown initially, the board is asymmetrical on the identical shafts. The shaft supporting the greater part of the board's weight will momentarily experience the greater frictional force, thereby enabling it to push the board toward the other shaft. When the situation reverses itself, the other shaft repays the favor, and the board returns to the first shaft. As long as the other shaft, through friction, can prevent the moving board from going too far—that is, prevent the board's center of mass from going beyond the shaft—then the oscillations will continue indefinitely, or until the board wears away completely.

*294. A Cranking Bicycle

The bicycle will move backward and the crank will rotate clockwise! Looking at the diagram, we note the following relationship: the resultant torque acting on the crank is zero—that is, $T r_2 - F r_1 = 0$, where T is the tension in the chain and F is the applied force. The resulting torques acting on the rear wheel is also zero—that is, $T r_3 - S r_4 = 0$, where S is the forward force exerted on the rear wheel by the ground. Newton's second law says that an acceleration should occur in the direction of the net force. If $S > F$, the net force is forward, so the acceleration is in the forward direction.

Combining the two equations and solving for S, we get $S = T r_3/r_4 = F r_1 r_3/(r_2 r_4)$. As shown in the diagram, $r_1 < r_4$ and $r_3 < r_2$ for any normal bicycle, so $S < F$. There is a net force $F - S$ acting *backward* on the bicycle frame. The bicycle will accelerate backward, with the wheels and the crank rotating clockwise.

Nightingale, J. D. "Which Way Will the Bike Move?" Physics Teacher *31 (1993): 244–245.*

*295. Turning a Corner

When you need to turn a corner on a bicycle, the curvature of the path could generate enough centrifugal force to knock us over to the outside of the turn. To counteract that, you lean into the turn so that the resultant force produced by gravity and the centrifugal force will lie in the tilted plane of the bicycle. To obtain the required tilt, you subconsciously ease the front wheel over to the outside of the turn. The resulting centrifugal force immediately tips us over toward the corner. To get out of the turn, you turn even more sharply into the bend, an action that throws the bicycle toward the vertical. The moment you are upright again, you simply straighten out the front wheel and once more follow a straight path.

In addition to all the above, there is an effect introduced by Starley in 1885, the bent fork, which plays a significant role in the stability of the bicycle during the turn. The result is a torque that twists the front wheel in the direction of the turn, precisely the effect needed to allow "no hands" riding.

But just as the bent-fork front wheel twists the front wheel into the turn, we must ask what effect prevents the front wheel from making too great an angle with the rest of the bicycle. The castering forces do just that! The principle is simple: A wheel carrying a load will roll easily in the direction in which it is pointing. The wheel will jam up, however, if you try to slide it sideways. The cure? Set the wheel axis slightly back from the swivel axis. The large force of sliding friction will soon align the wheel in the direction of motion. The bicycle behaves in a similar fashion. The rider and the rest of the bicycle swivel behind the front wheel, which defines the direction of motion. To the bicyclist, however, it will appear that it is the front wheel that exhibits the self-centering action.

Kirshner, D. "Some Nonexplanations of Bicycle Stability." American Journal of Physics *48 (1980): 36–38.*

*296. Race Driver

Of course, the driver slows down enough *before turning* so that the car will not be thrown off the road when the driver speeds up around the turn. The goal is to have the higher exit speed into the straightaway.

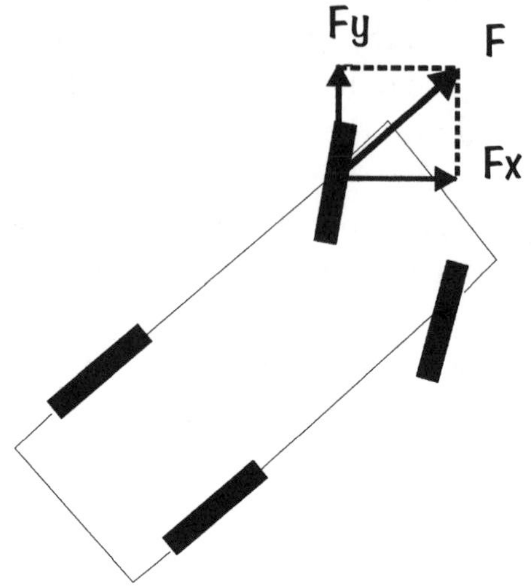

As shown in the diagram, when the front-wheel-drive car accelerates while turning, the front wheels receive an extra push, represented by the force *F*. This extra force *F*, when resolved into its components with respect to the car body's forward and sideways directions, can be seen to push the car forward out of the turn and sideward into the turn. If either force component exceeds the static frictional force maximum value between the tires and the roadway, slippage will occur.

Franklin, G. B. "The Late Apex Turn." Physics Teacher *28 (1990): 68.*

Hewko, R. A. D. "The Racing Car Turn." Physics Teacher *26 (1988): 436–437.*

*297. The Wall Ahead

The kinetic energy of the car must go to zero to bring the car to rest before hitting the wall. If the stopping distance x is less than the distance d to the wall, no collision occurs. When steering straight toward the wall and braking, the "work done" by the static frictional force F (no sliding) acting over the stopping distance x is determined from $mv^2/2 - Fx = 0$. Solving for the stopping distance $x = mv^2/2F$.

For the circular turn with no braking, the centrifugal force acts so that when no sliding occurs, $F - mv^2/R = 0$, or $R = mv^2/F$. As long as x or R is less than the distance to the wall d, the car will not collide with the wall. One can see at once that, for the same static frictional force F, the stopping distance x is $R/2$. So you are better off braking while going toward the wall!

Chapter 10 Born to Run

298. Strong Woman

True. Numerous studies report that per kilogram of lean body weight, females are actually somewhat stronger than males. These results imply that a female need not possess quite as much muscle to do as well as a male in the same event, all other factors being equal. In addition, athletes typically use no more than 20 percent of their muscle potential. So the fact that today some fourteen-year-old girls swim faster than Johnny Weissmuller—the original Tarzan—did in the 1924 Olympics should come as no surprise. Women can develop substantial strength through weight training without building up bulging muscles. If the man is stronger, he simply has more muscle tissue than the woman. The woman's body will tend to have more body fat by weight, about 25 percent, whereas the man's body will be about 15 percent for the same overall conditioning.

Alexander, R. McNeill. The Human Machine. *New York: Columbia University Press, 1992, pp. 35–39.*

Wilmore, J. H. "A Look at the Female Athlete Proves a Woman Is Not Limited by Her Biology." Women in Sports *June (1974): 40.*

299. Hanging in the Air!

Hanging in the air is an illusion that is easily explained by basic physics. As the body reaches the peak of the jump, the vertical speed is very low, so the body does not move very far. One can calculate that the person spends half of the time in the air at the top quarter of the trajectory. The actual time period for "hanging in the air" is less than one second, but the slower motion fools our eye-brain system into thinking otherwise!

300. Good Running Shoes

Runners need good shoes for maximum comfort and for better running results. The good running shoe performs two major functions: it provides friction with the ground to prevent slipping forward or backward; and it adds additional springiness, with the shoe ideally acting as an extension of the Achilles tendon. Any failure to accomplish either of these functions well means that some of the runner's energy does not move the legs and body in the desired manner, or in the vernacular, some energy is wasted.

As the front foot lands, good shoes not only compress over the desired extended time interval but also are able to spring back into shape during the desired recovery time interval. The timing, the amount of compression, and the location of the compression are all important parameters making optimal shoe design difficult. For example, different distances for racing require special refinements. A sprinter runs mostly on the balls of her feet, so more cushioning is needed there than in the heels or in the middle of the feet. Middle-distance runners have a midfoot-to-toe stride (or a less efficient heel-to-toe stride), which requires more spring from heel to midfoot. Shoe designs have improved, but there still is room for improvement.

Olympics Editor. "Running Shoes." Newsweek *(July 27, 1992): 58.*

301. Sprinting

Chemical energy is made available in the muscle cells by two mechanisms: aerobic (with oxygen) and anaerobic (without oxygen). For races under ten seconds, there isn't enough time for the oxygen inhaled *during the race* to contribute to the conversion of chemical energy for the muscles. The oxygen already inhaled before the 100-meter race begins does contribute to the total energy requirement, which is about 7 percent aerobic and 93 percent anaerobic.

Frohlich, C., ed. Physics of Sports. *College Park, Md.: American Association of Physics Teachers, 1986, pp. 113–123.*

Ward-Smith, A. J. "A Mathematical Theory of Running, Based on the First Law of Thermodynamics, and Its Application to the Performance of World-Class Athletes." Journal of Biomechanics *18 (1985): 337–349.*

302. Long-Distance Running Strategy

The racers want to avoid excessive exercise during the early stages of the race so that lactic acid buildup in the muscles, the product of the glycolytic anaerobic mechanism, is delayed until the final stages of the race. The lactic acid in the muscles leads to discomfort and to diminished levels of performance.

Frohlich, C., ed. Physics of Sports. *College Park, Md.: American Association of Physics Teachers, 1986, pp. 113–123.*

Strnad, J. "Physics of Long-Distance Running." American Journal of Physics *53 (1985): 371–373.*

Ward-Smith, A. J. "A Mathematical Theory of Running, Based on the First Law of Thermodynamics, and Its Application to the Performance of World-Class Athletes." Journal of Biomechanics *18 (1985): 337–349.*

303. Location Effects on High-Jump Records

One major reason why the variation in *g* is ignored when world records are considered for the high jump and the long jump is the fact that other parameters play a much more important role. A small breeze within the allowable limit of 2 meters per second, or the condition of the grass and the soil on the approach, or the temperature and humidity, or the air density, or the bend in the bar, all can vary within certain values and have a greater effect on the athlete's result.

An extreme in the difference of *g* between two Olympic sites, Mexico City and Moscow, is very small, at about 0.4 percent, while the difference in air density is 22.2 percent. In the high jump, compensation of these two effects results in a 3-millimeter difference, insignificant compared to the nearest centimeter, at which high-jump heights are measured. In the long jump, however, the difference is nearly 5 centimeters, about half due to decreased *g* value at Mexico City and the other half due to decreased air resistance. (Even after adjustment, Bob Beamon's long-jump record at Mexico City in 1968 would have remained a world record performance until 1991, when it was beaten near sea level in Tokyo by Mike Powell of the United States, who jumped 8.96 meters.)

Ficken, G. W. Jr. "More on Olympic Records and g." American Journal of Physics *54 (1986): 1063.*

Frohlich, C. "Effect of Wind and Altitude on Record Performance in Foot Races, Pole Vault, and Long Jump." American Journal of Physics *53 (1985): 726.*

Kirkpatrick, P. "Bad Physics in Athletic Measurements." American Journal of Physics *12 (1944): 7.*

McFarland, E. "How Olympic Records Depend on Location." American Journal of Physics *54 (1986): 513.*

304. High-Jump Contortionist

At the higher heights, this Fosbury flop technique is the only way by which jumpers can clear the bar. Even the best athletes can raise the center of gravity only about 80 centimeters (2 feet, 7 inches). If the jumper's center of gravity begins at 1.1 meters above the ground (for a tall jumper), the highest altitude attainable for the center of gravity will be $1.1\ m + 0.8\ m = 1.9\ m$, or about 6 feet, 4 inches above the ground. In executing an 8-foot jump, the jumper's center of gravity will pass 1 foot, 8 inches below the bar! Therefore, the center of gravity is outside the body of the jumper when the jumper is contorted this way.

305. Pole Vaulter

First, one should definitely have the best pole (i.e., greatest elasticity), so that more of the energy transferred into the pole during the initial bending is transferred back into lifting the vaulter and the pole during the vault. But how long should the pole be? That is the question. A longer pole adds weight, which will result in a slower speed just before planting the pole in the box. The speed squared is proportional to the kinetic energy of the pole vaulter system as it approaches the vault, and a slower speed produces less bending of the pole and less energy available to be transferred back into the upward lifting by the pole.

Adding to the limitation is the requirement that the vaulter's forward horizontal motion near the top must be able to move the body horizontally over the crossbar. With a longer pole *and* a grip farther back than before, the running speed must be great enough to bend the pole sufficiently to allow the pole while in the process of straightening to accomplish this horizontal movement of the vaulter with the correct timing. If the running speed is insufficient, the vaulter will not move forward ahead of the planted end of the pole in the vault box as the pole unbends. Each pole vaulter attempts to maximize his vault height by improving technique through a combination of running speed, grip position, body position changes, and pole selection.

306. Basketball

When the ball swishes through the net without hitting the rim, then the backspin technique only contributes to the accuracy of the shot as far as distance and entry angle are concerned. The greatest value of backspin occurs when the ball does not swish through the net, either directly or off the backboard. But if the basketball hits the rim, backspin will produce a minimum translational distance afterward as well as reduced spinning afterward. Physics analysis reveals that "a backspinning ball always experiences a greater decrease in translational energy and in total energy than a forward-spinning ball." Hence the shot seems "softer" and more likely to drop into the basket after hitting the rim.

Brancazio, P. J. "Physics of Basketball." American Journal of Physics *49 (1982): 356–365.*

Erratum. "Physics of Basketball." American Journal of Physics *50 (1982): 567.*

307. Doing the Impossible!

To rise on tiptoes, you must shift your weight forward, but the doorframe edge prevents your forward movement. There is a way to accomplish this feat, although supplementary objects are required. Grab two heavy objects (books will do), assume the prescribed position at the doorway edge, swing your arms forward, and rise on tiptoes.

308. Reaction Time with a Bat

Some of the best professional batters say that they begin their swing after the pitcher releases the ball, while others say they can wait just long enough to see the ball spin a few feet out of the pitcher's hand. Most batters require a few tenths of a second for the swing, which, for a ball coming at 90 miles per hour, means that the swing must definitely begin when the ball is at least 20 feet away.

Most amateur batters should begin their swing of the bat just after the release of the ball by the pitcher! Otherwise, one soon discovers that the bat crosses home plate after the ball hits the catcher's mitt, unless the batter possesses great wrists to provide fast bat speed. Just step up to the plate in a pitching cage where the pitch travels at 80 miles per hour or more to experience this fate.

309. Can Baseballs Suddenly Change Direction?

Yes. In fact, about 75 percent of the total deflection occurs during the last half of the flight, and a whopping 50 percent can occur during the last few feet of the flight! How? Take the simplest case to analyze—the constant acceleration case. If the acceleration effect of the air on the spinning ball is taken as a constant (for simplicity), then one has the familiar expression $s = 1/2\ at^2$ for the displacement distance s, constant acceleration value a, and clock time t. Hence the amount of displacement goes as the time interval squared. As an example, let the displacement for the first half of the flight be 1 inch. Then the total displacement for the total flight will be 4 inches.

In a more representative case, the acceleration from the air due to the spin of the ball can be even more pronounced near the plate than calculated for the constant acceleration case above.

310. The Curveball

Thrown properly, the curveball thrown by a right-handed pitcher curves *downward* mostly, with some additional movement to the left, away from a right-handed batter. The curveball moves downward and to the right from a left-handed pitcher. The pitcher puts topspin on the ball, with angular momentum components in two directions: spin about the horizontal axis with the ball turning over the top from back to front, and a little bit of spin about the vertical axis with the ball turning counterclockwise when seen from above. In most cases, some spin about the other horizontal axis is imparted also.

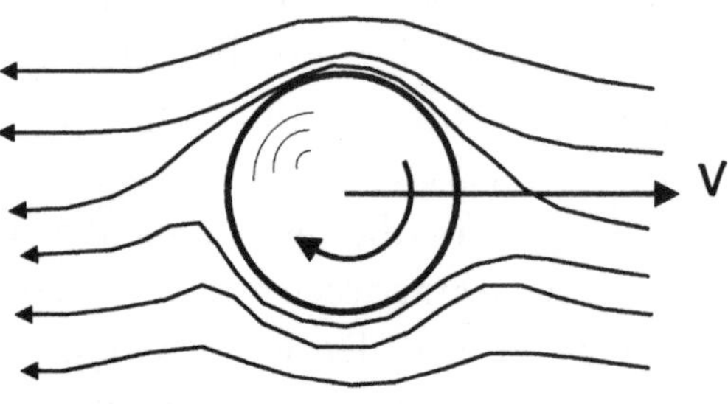

The explanation for the curved path produced by the spin alone lies with the Magnus effect, an application of the Bernoulli principle. The spinning baseball causes a very thin layer of air, called the boundary layer, next to its surface to rotate with the ball rotation. The spinning ball moving through the air affects the manner in which the general air flow separates from the surface in the rear and in turn affects the general flow field about the body. Consequently the Magnus effect arises when the flow follows farther around the curved surface on the side traveling with the wind than on the side traveling against the wind in the same time interval. The air flow on the top side of the baseball is slightly slower and on the bottom side is slightly faster. Bernoulli's principle tells us that there will be a net force downward, and the ball responds. The spin about the vertical axis creates a lower pressure on the left than on the right, so the ball moves leftward, away from the right-handed batter. For speeds up to 150 feet per second (about 100 miles per hour) and spins up to 1,800 revolutions per minute, the lateral deflection is directly proportional to the first power of the spin and to the square of the wind speed.

Adair, R. K. The Physics of Baseball. *New York: HarperCollins, Harper Perennial, 1990.*

Allman, W. F. "The Untold Physics of the Curveball." In Newton at the Bat: The Science in Sports, *edited by E. W. Schrier and W. F. Allman. New York: Charles Scribner's Sons, 1987, pp. 3–14.*

Briggs, L. J. "Effect of Spin and Speed on the Lateral Deflection (Curve) of a Baseball; and the Magnus Effect for Smooth Spheres." American Journal of Physics 27 *(1959): 589–596. Repr., A. Armenti Jr., ed.,* The Physics of Sports, vol. 1. *New York: American Institute of Physics, 1992, pp. 47–54.*

Watts, R. G., and A. T. Bahill. Keep Your Eye on the Ball. *New York: W. H. Freeman, 1990.*

311. Scuffing the Baseball

Scuffing a baseball is a prohibited act that gives the pitcher a definite advantage. Using a bottlecap, belt buckle, sandpaper, or whatever object the pitcher can sneak to the mound, the pitcher scuffs (roughens up the surface of) one spot on the ball. The ball is then thrown so that the scuffed spot is on the axis of rotation of the ball. The scuffed spot acts to delay the separation of the airflow, and the net force from the Bernoulli principle application will be toward the scuffed side. The additional force can increase the lateral force amount by as much as 30 percent or more! The baseball's flight path can certainly change more dramatically if desired.

Watts, R. G., and A. T. Bahill. Keep Your Eye on the Ball. *New York: W. H. Freeman, 1990, p. 75.*

312. Watching the Pitch

Although the batting instructor tells you to "keep your eye on the ball," not even professional baseball players can follow the pitched baseball traveling faster than 60 miles per hour (27 meters per second) to a point closer than 5 feet from the plate. To do so, one would need to turn the head at an angular speed of about 500 degrees per second—much too fast for humans to track. One can certainly anticipate by looking ahead of the ball to watch the ball strike the bat, and some ball players admit that they do this act occasionally.

Watts, R. G., and A. T. Bahill. Keep Your Eye on the Ball. *New York: W. H. Freeman, 1990, pp. 153–168.*

313. The Bat Hits the Baseball

No. Empirical measurements of wooden and aluminum bats show that the location along the bat that imparts the greatest speed to the hit baseball does *not* occur at the center of percussion position. The best response occurs at the maximum energy transfer (MET) point, which also lies beyond the center of mass for practically all bats.

For bats of identical shape, the aluminum bat has a slightly wider region representing high batted-ball speeds than does the wooden bat, and this region is skewed a bit more toward the handle. Batters have reported that aluminum bats allow them to hit inside pitches harder, meaning that these balls go farther than they would when hit with a wooden bat.

Watts, R. G., and A. T. Bahill. Keep Your Eye on the Ball. *New York: W. H. Freeman, 1990, pp. 124–125.*

314. Underwater Breathing

The water pressure at a depth of 2 meters would make breathing through a tube impossible for any length of time, and even a muscular person would find taking a few breaths to be very strenuous. The overwhelming forces are produced by hydrostatic pressure, which is often forgotten until one experiences it underwater.

315. Springboard Diving Tricks

There is no need to begin *both* the twisting and the somersaulting before leaving the springboard. What is required is some nonzero angular momentum about a body axis before beginning the second rotation type. Typically there is a small amount of forward rotation as the diver leaves the board, with the angular velocity vector parallel to the angular momentum vector. The diver can speed up the rotation by moving into a tuck position, keeping the two vectors parallel. Or the diver can begin a twisting rotation by moving one arm above the head and the other downward, across the body. In this case, the body will respond by tilting from the vertical slightly to keep the total angular momentum vector from both rotations identical to the initial value and direction, since no external torque is being applied. Note that the angular momentum vector and the angular velocity vector are no longer parallel now, but the reason can be traced back to unequal moments of inertia about two perpendicular body axes and the fact that the moments of inertia can be changed.

Frohlich, C. "Do Springboard Divers Violate Angular Momentum Conservation?" American Journal of Physics *47 (1979): 583–592. Repr., A. Armenti Jr., ed.,* The Physics of Sports, vol. 1. *New York: American Institute of Physics, 1992, pp. 311–320.*

———. *"The Physics of Somersaulting and Twisting."* Scientific American *259 (1980): 155–164.*

316. Cat Tricks

The drawings are tracings from a film at roughly 1/20-second intervals, showing eight consecutive positions of a cat during its descent. There are no external torques acting on the cat, so its net angular momentum about any axis must remain constant throughout the fall. In fact, the angular momentum about any axis must be zero if the cat was simply dropped without any rotational motion.

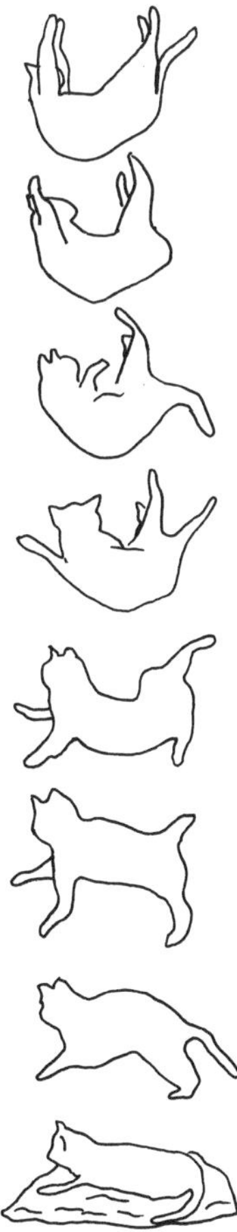

The cat's behavior can be understood by thinking about the cat as consisting of two halves—the front half and the rear half. The drawings show that the front half of the cat is righted first. After first drawing in its front paws to diminish the moment of inertia about the long body axis for the front half, the cat extends the hind limbs to increase the moment of inertia for the rear half about the body axis. The cat then rotates the front half through at least 180 degrees, with the rear half rotating in the opposite direction through a much smaller angle.

Once the front half is righted, the haunches are swung around by drawing in the hind limbs and extending the front paws, in contrast to the first stage. Rotation of the rear half now occurs, with the front half rotating back slightly. A vigorous rotation of the tail helps, but even tailless cats can right themselves before landing.

Essén, H. "The Cat Landing on Its Feet Revisited, or Angular Momentum Conservation and Torque-Free Rotations of Nonrigid Mechanical Systems." American Journal of Physics *49 (1981): 756–758.*

Fredrickson, J. E. "The Tailless Cat in Free-Fall." Physics Teacher *27 (1989): 620–621.*

Kane, T., and M. P. Scher. "A Dynamical Explanation of the Falling Cat Phenomenon." International Journal of Solids Structure *5 (1969): 663.*

317. Astronaut Astrobatics

Yes. Just like the diver and the cat, an astronaut can initiate rotation about any chosen axis. However, the body needs to have some movement—say, torso movement relative to leg movement. One does "tuck drops" to rotate about a somersaulting axis and "swivel hips" to rotate about a twisting axis.

Frohlich, C. "Do Springboard Divers Violate Angular Momentum Conservation?" American Journal of Physics *47 (1979): 583–592. Repr., A. Armenti Jr., ed.,* The Physics of Sports, vol. 1. *New York: American Institute of Physics, 1992, pp. 311–320.*

———. "The Physics of Somersaulting and Twisting." Scientific American *259 (1980): 155–164.*

318. The Feel of the Golf Shot

Yes and no, for the ball has left the club head before the hand-brain system feels the blow! One can calculate the travel time up the club shaft for the sound wave: assume a 3-foot distance at about 15,000 feet per second, and the delay is 0.0002 second. But the sensation must go to the brain to be "felt," an additional delay of up to 15 to 20 milliseconds. The golf ball contact time is usually fewer than 10 milliseconds, so the sensation is felt *after* the ball has left the club head.

319. Skiing Speed Record

The skiing record downhill is about 2 percent faster than the terminal speed for falling through the air because the skier can use his poles to apply an additional force. Skiers going down Mount Fuji in Japan are famous for their enormous speeds on the slopes!

320. "Skiers, Lean Forward!"

For the skier, the body should be aligned along the local "up" direction. If the snow were frictionless, this "up" direction is perpendicular to the slope. If the skier accelerating downhill on the frictionless snow happens to be carrying a simple plumb bob on a string, the string's rest position would be perpendicular to the slope. If the skier tries to remain vertical—that is parallel to the trees—the skis will slip out from underneath.

When the wind effect increases with increasing speed, the person will want to lean forward even more to avoid being blown over.

Bartlett, A. A., and P. G. Hewitt. "Why the Ski Instructor Says, 'Lean Forward!'" Physics Teacher *25 (1987): 28–31.*

321. Ski Slope Anticipation

Suppose the skier enters a short region where the slope of the ski run changes abruptly, by 5 degrees or so. Without the "prejumping" technique, the skier will leave the ground for about 1/2 second, and she will feel a vertical force on her legs upon impact of up to several times her body weight. Such a large impact force could affect her stability.

Prejumping minimizes the impact from the landing force by attempting to land the skier immediately at the beginning of the steeper slope and parallel to the slope. By raising her skis off the snow at the correct distance before the steeper slope is encountered, the skier's body begins to fall, and the skis can make contact almost immediately on the steeper slope with a much smaller impact force upon landing. Of course, the skier also must learn to rotate the ski tips downward through a small angle in order to land parallel.

Hignell, R., and C. Terry. "Why Do Downhill Racers Prejump?" Physics Teacher *23 (1985): 487–488.*

Swinson, D. B. "Physics and Skiing." Physics Teacher *30 (1992): 458–463.*

322. Riding a Bicycle

Examining the body motion details for running and bicycle pedaling can become quite complicated. So we attempt a reasonably rough approximation that retains the essential factors; assume that the legs experience identical movements for both cases. (One would expect the bicycle rider's legs to move less to cover the same distance.) During the running, the legs move up and down, and the torso moves up and down. During the bicycle riding, the torso remains fixed vertically, but the legs move up and down to match the runner's leg movements. The runner must do additional work to move the torso vertically. *Voilà!*

The extra perspiration and heating during running remind us that the physiological system knows the laws of physics, too. By measuring the oxygen requirements, exercise physiologists have computed the energy needs to be about 260 kilojoules per kilometer of running for a 700 newton person (about 160 pounds), and the energy needs are considerably less for bicycle riding.

DiLavore, P. "Why Is It Easier to Ride a Bicycle than to Run the Same Distance?" Physics Teacher 19 *(1981): 194.*

323. Give Me a Big V

Yes. Each bird pushing downward with its wings on the air below creates an updraft around it. If other birds crowd in close, they can use those updrafts to help keep themselves aloft. Only the lead bird cannot take advantage of this updraft. Calculations reveal that a flock of twenty-five birds can fly in formation some 70 percent farther than one bird alone.

324. Deadly Surface Tension

A person coming out of the bath or shower may be carrying a thin film of water that weighs roughly 1 pound (0.5 kilogram). A wet mouse would be carrying about its own weight in water! A wet fly would be lifting many times its own weight in water, and once wetted by the water is in great danger of remaining so until it drowns. These consequences are the result of the surface-to-volume ratio, which is very large for tiny insects and very small for large animals.

*325. Animal Running Speeds

The power developed by an animal is proportional to the cross-sectional area L^2 of its muscles because its strength is proportional to L^2, where L is the animal's linear size. On level ground, the power is needed in overcoming air resistance, an opposing force proportional to the animal's cross-sectional area and the square of its speed v. Therefore, $F_{air} \propto L^2v^2$, and the power of the air resistance is $P_{air} = F_{air}\, v \propto L^2v^3$. Setting the power generated equal to the power needed, one learns that the speed v is independent of L.

Running uphill involves slower speeds, so one can neglect the air resistance power term compared to the rate of change of gravitational potential energy, which is proportional to mgv. But m is proportional to L^3, so the rate of change of potential energy goes as L^3v. Now one finds that $v \propto 1/L$. Thus smaller animals can run uphill faster than larger ones.

*326. Scaling Laws for All Organisms

One would expect that the energy requirements should grow as the first power of the body mass, but the empirical results give the body mass to the power 3/4. So the explanation must lie in how the needed resources are distributed within the body. When the following three conditions are met, the capillaries and arteries of the circulatory system make the heart work no harder than necessary to deliver blood throughout the body.

1. The delivery system, to reach every part of an organism, must be a branching, fractal-like network that fills the whole body.

2. The terminal branches of this network are the same size in all organisms.
3. Evolution has tuned the networks to minimize the energy required to deliver the goods.

Several other established power laws for other biological properties also follow in this model, such as the slower breathing for larger animals, empirically giving a respiratory rate inversely proportional to the body mass raised to the power 1/4.

McMahon, T. "Size and Shape in Biology." Science *17 (1973): 1201–1204.*

West, G.; J. Brown; and B. Enquist, as reported by R. Pool. "Why Nature Loves Economies of Scale." New Scientist *(April 1997): 16.*

*327. Tennis Racket "Sweet Spot"

There are actually three "sweet spots" on the face of a tennis racket, each based upon a different physics principle. When the ball strikes any of the sweet spots, the stroke will feel good for different reasons. So far, no one has been able to make a tennis racket with all three sweet spots at the same location, although some of the larger rackets have moved them much closer together.

The first sweet spot is at the node of the first vibrational harmonic. When the ball strikes the racket, the fundamental vibrational mode is at about 30 hertz, and its harmonics are excited. The first harmonic is about 150 hertz, with its node on the central axis, slightly above the center of the strings. When the ball strikes this node, the significant decrease in vibration is noticed by the player.

The second sweet spot is at the center of percussion, so the ball striking here will not attempt to rotate the racket. The player feels no twisting force at the hand. This twisting sweet spot is about 2 inches below the center of the strings.

The third sweet spot is called the point of maximum coefficient of restitution (COR). A tennis ball striking here maintains more of its initial kinetic energy. Tighter strings will cause more deformation of the ball on impact, with less kinetic energy after the collision. One way to increase the COR of a racket is to string it with less tension. The point of maximum COR is about 1 inch above the bottom edge of the strings.

Brady, H. "Physics of a Tennis Racket." American Journal of Physics *47 (1981): 816.*

*328. Golf Ball Dimples

The dimples play two roles. They cause the drag force to decrease suddenly at velocities above approximately 25 meters per second (82 feet per second), being about half the drag experienced by a smooth sphere. The dimples also directly affect the aerodynamic lift. Various patterns of dimples are available, and some of the latest patents include two sizes of dimples covering more than 79 percent of the surface.

While rough golf balls do, paradoxically, experience less air resistance, the primary purpose of dimpling is to increase the lifting force on a ball, given bottom spin. How does roughness reduce drag? At low speeds it does not; but a full drive sends a golf ball flying at 160 miles per hour (250 kilometers per hour). A ball flying through the air is enveloped by a thin boundary layer. If the ball is smooth, the boundary layer is laminar—that is, there is no mixing of the sublayers. The main flow separates from the ball, producing a region of backflow and large eddies downstream. But if the ball is rough, the air in the boundary layer must go over the hills and valleys. The flow becomes turbulent, which means a lot of mixing and momentum exchange. As a result, the high-speed air flowing outside the boundary layer is able to lend momentum to the low-speed air inside the boundary layer. With this assistance the turbulent boundary layer can flow farther against increasing pressure than the laminar boundary layer can. The main flow remains attached to the ball, making the low-pressure eddying region on the downstream side much smaller than in the laminar case. Moreover, the pressure on the downstream side is not as low. Therefore the force imbalance between the downstream side and the upstream side of the ball is reduced. That is, the form drag is less.

The dimples create lift. The ball can impart a spinning motion to only a thin layer of air. In addition, the laminar boundary layer does not follow all the way around the ball. Instead, the boundary layer separates earlier on the side spinning against the relative wind, the bottom side for the golf ball. A turbulent boundary layer can exchange momentum with the relative wind much more than a laminar boundary layer can. As a result, there will be lift.

Erlichson, H. "Measuring Projectile Range with Drag and Lift, with Particular Application to Golf." American Journal of Physics *51 (1983): 357–362.*

MacDonald, W. M., and S. Hanzely. "The Physics of the Drive in Golf." American Journal of Physics *59 (1991): 213–218.*

Chapter II
Third Stone from the Sun

329. California Cool

The cooler California coast is the result of the Coriolis force, which makes everything in the Northern Hemisphere sidle to the right of its motion. The prevailing winds that drive the water onto the California coast are from the northwest, which means that the Coriolis force transports water away from the shore toward the southwest. The resulting deficit is made up by cold water rising from depths of several hundred feet and forming a cool strip of water along the coast. In addition, the cold California current flows down from the north and lowers the temperature of the coastal waters even more.

330. Waves at the Beach

The inshore part of each wave is moving in shallower water, where the friction of the bottom causes the wave to slow down. Thus the inshore part moves slower than the part in deeper water. The result is that the wave front tends to become parallel to the shoreline. We can also see that this process has the effect of concentrating wave energy against headlands. It is a modern expression of the old sailor's saying "The points draw the waves."

Bascom, W. Waves and Beaches: The Dynamics of the Ocean Surface. *Garden City, N.Y.: Doubleday, Anchor Books, 1964, pp. 70–77.*

331. Ocean Colors

The reflection coefficient for light reflected from the surface of water decreases when the angle of incidence (measured with respect to the vertical) becomes smaller. When looking straight down, you receive rays reflected at very small angles. The rays reflected from the water surface near the horizon are at larger incidence angles from the perpendicular, so fewer of them are absorbed by the water.

332. Stability of a Ship

A stable ship is one that can right itself if it is heeled over. As seen in the diagram, the ship's center of buoyancy *B* must move in the direction of the slope so its upward push (and its counterclockwise torque) can combine with the downward force attached to *G*, the ship's center of gravity. Only then can the ship right itself. The ship's stability is measured by the distance *GM* between G and the so-called metacenter, a point at the intersection of the centerline of the hull and a vertical line through B. A safe GM for the average fully loaded merchant ship is about 5 percent of her beam—the breadth at the widest part.

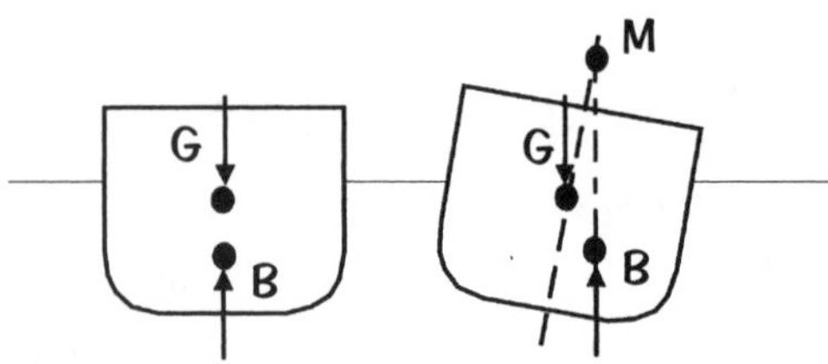

333. Longer Ships Travel Faster

A surface ship creates waves as it moves, including a bow wave in front and additional waves along its length and at its stern. At hull speed, the ship is left with a bow wave and a stern wave, the two separated by the length of the ship's hull. Here it is important to remember that in deep water longer waves travel faster (i.e., $v = \sqrt{g\lambda/2\pi}$). A ship trying to exceed its hull speed will have to cut through or climb up its bow wave. At that point the ship's power requirements rise steeply, and getting ahead becomes an uphill battle.

Vogel, S. "Exposing Life's Limits with Dimensionless Numbers." Physics Today *51 (1998): 22–27.*

334. Polar Ice

Antarctica is a continent. Land is a poor heat conserver, radiating heat away as soon as it gets it. (This behavior indicates why winters are harsh deep inland.) The Arctic ice is over an ocean, and water is known for its high heat capacity, taking a long time to heat but once it is warm, losing heat slowly. The Arctic stores summer heat and lives off its "savings" in the winter.

335. The Arctic Sun

The direction the observer was facing can be deduced by examining the situation at two other latitudes. At the North Pole, the elevation of the sun would be nearly constant during the day. At latitudes between 30 degrees and 45 degrees, the sun reaches its highest elevation when it is due south, and its lowest upon rising in the east and setting in the west. As we move farther north, we expect the rising and setting locations to move farther north until they meet due north of us. Therefore the observer was facing north.

The sun reaches its highest elevation when it is due south. This time is known as local noon. The lowest elevation is then obtained at local midnight.

336. Circling Near the Poles

The effect may be due to the Coriolis force, which is about 50 percent stronger at the poles than in middle latitudes. When walking we make corrections for the Coriolis force easily and quite unconsciously. On near-frictionless ice this is impossible. A person somehow able to walk at 4 miles per hour on near-frictionless ice would drift from her intended straight path by about 250 feet at the end of a mile. One occasionally hears stories that even the penguins in the Antarctic waddle in arcs to the left, although the authors cannot guarantee the scientific accuracy of this statement.

McDonald, J. E. "The Coriolis Force." Scientific American *72 (1952): 186.*

337. Weather Potpourri

They are all true!

1. A rainstorm occurs in an area of low barometric pressure. When there is less air pressure on your body, the gases in your joints expand and cause pain.
2. A storm is often preceded by humid air. Frogs have to keep their skins wet to be comfortable, and moist air allows them to stay out of the water and croak longer.
3. A low-pressure rain system moving into an area will often stir up a south wind that flips leaves over.
4. Ice crystals form in high-altitude cirrus clouds that precede a rainstorm. These crystals refract light from the moon and make a ring around it.
5. Birds' and bats' ears are very sensitive to air pressure changes. The lower pressure of a storm front would cause them pain if they flew higher, where the pressure is even lower.
6. Cold-blooded crickets chirp more the hotter it gets. Count the number of chirps a cricket makes in 15 seconds and add 37—this number will give you the temperature in degrees Fahrenheit.
7. Rising humidity causes ropes to absorb more moisture from the air, and this process makes them shrink.
8. Fish come up for insects that are flying closer to the water before a storm because of lowered atmospheric pressure.
9. A rising wind, often marking the coming of a storm, causes a high whining sound when it blows across telephone wires.

338. Wind Directions

False! If winds rushed directly toward areas of lower pressure, no strong "highs" or "lows" could develop, and our weather would be much less changeable than it is. Instead, due to the Coriolis force caused by the rotation of the earth, wind from any direction veers to the right in the Northern Hemisphere. As a result, the whole air mass initially flowing directly toward a low-pressure area begins to rotate counterclockwise. This rotation in turn prevents the filling of the low-pressure region, since now the pressure difference supplies a centrifugal force that tends to keep the winds moving in circular paths. In the Southern Hemisphere the Coriolis force causes winds to veer to the left, and so the direction of circulation is clockwise.

Near the Equator, the Coriolis force is zero or very small. In that region any atmospheric pressure differences produced by heating of the air at the ground are quickly smoothed out, and the region has well earned the name of "the doldrums." Hurricanes and typhoons rarely form closer to the Equator than 5 degrees latitude.

339. Deep Freeze

The astronomical reason is the earth's elliptical orbit. At perihelion, the point of the orbit nearest the sun, the earth is 1.407×10 kilometer from the sun. At aphelion, the point farthest from the sun, the earth-sun distance is 1.521×10 kilometer. The difference is relatively small, but it is not negligible. Happily for the Northern Hemisphere, perihelion occurs during winter on January 4 or 5, and this timing helps to moderate the seasonal effect produced by the tilt of the earth's axis to the orbital plane.

The reverse is true for the Southern Hemisphere, which would suggest that the latter should have colder winters and hotter summers. However, the greater area of ocean south of the Equator serves as a moderating influence. The high heat capacity of water means that in summer the ocean is slow to warm and in winter slow to cool. This physical property makes summers in the Southern Hemisphere somewhat less hot and winters somewhat less cold than they would be otherwise.

340. Weather Fronts

Near the ground, the higher-air-pressure regions are generally cold and the lower-pressure regions warm. However, for higher altitudes we must consider the variation of pressure and density with height. Due to gravity, most of the atmosphere is concentrated near the ground. The reason why all of the atmosphere does not collapse completely is that the downward gravitational pull on each parcel of air is balanced by the upward push due to the higher pressure from below. This force balancing happens if the pressure and density of the atmosphere decrease exponentially upward. The exact formula is $P = P_0$ $exp(-mgh/RT)$, where h is the height and P_0 is the pressure at ground level. We see that pressure decreases with height more slowly in warm air than in cold air (see the diagram). As a result, at any given height the pressure is higher in the warm zone than in the cold zone.

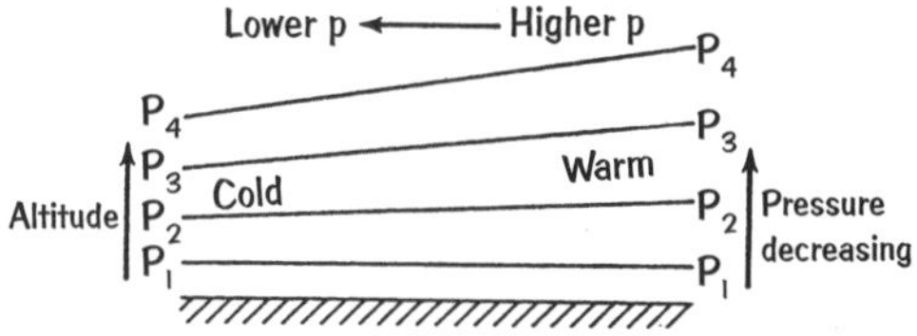

This horizontal pressure difference grows with height and generates the thermal wind. For example, the thermal wind associated with the polar-subtropical temperature difference is always on average westerly and manifests itself as the circumpolar jet stream, which snakes around the pole in a wavy manner.

341. Lightning and Thunder

The main reason for the rumbling, clapping, and other sounds is that lightning follows a sinuous path. Some points on its path will lie closer to the observer than others, so the sound of thunder will be extended. If the nearest point is 5,000 feet closer than the farthest point, the thunder will roll about 5 seconds, since the speed of sound in air is about 1,000 feet per second. Also, lightning often consists of many strokes following each other in rapid succession. Thirty to forty strokes have been observed along much the same path at 0.05-second intervals. The sound waves produced by multiple lightnings interfere with one another, resulting in thunder that intensifies and diminishes

Most of the acoustic energy is radiated perpendicularly to a segment of the lightning channel. Hence, if the entire channel is oriented roughly at right angles to the observer's line of sight, a much larger portion of the radiated energy will be received. Equally important, all points in the channel will produce sound that arrives almost simultaneously at the observer, and the result is a high-intensity sound—a peal or a clap. The pitch of thunder depends primarily on the energy of the lightning stroke. The more powerful the stroke, the lower the pitch. A typical value is 60 hertz.

Few, A. A. "Thunder." Scientific American *233 (1975): 88–90.*

342. Lightning without Thunder?

Strictly speaking, no. But there may be lightning whose thunder is inaudible even a fairly short distance from the lightning channel. For example, there have been lightning flashes that reportedly struck the Washington Monument without producing thunder audible to people nearby.

If there is no return stroke and the flash consists of only a low-level current, as occasionally happens in structure-initiated flashes that move upward from the building tops, one can expect very little sound to be generated.

Uman, M. A. All about Lightning. *New York: Dover Publications, 1986, pp. 113–115.*

343. Direction of the Lightning Stroke

In a sense, lightning does both, going both up and down a lightning channel. A cloud-to-ground discharge begins in the form of a stepped leader, a faint downward-moving traveling spark that follows a highly irregular series of steps, each about 50 meters long. When the leader gets to within roughly 100 meters of the ground, sparks will be emitted from the objects and structures on the ground, typically from the highest points first. One of these upward-going discharges contacts the leader, thereby determining the point where the lightning will strike. When the leader is attached to the ground, the return stroke begins, in which the electrons at the bottom of the channel move violently to the ground, causing the channel near the ground to become very luminous. Then in succession the electrons from higher and higher sections of the channel flow toward the ground, reaching currents of about 20,000 amperes and sometimes as high as 200,000 amperes. The channel expands at supersonic speed to a luminous diameter of perhaps 5 or 6 centimeters. The stepped leader may require 20 milliseconds to create the channel to the ground, but the return stroke is completed in a few tens of microseconds. Typically the process is repeated three or four times, utilizing the old channel to produce a lightning flash with an average duration of 0.2 second.

To summarize, electrons at all points in the channel usually move downward, even though the region of high current and high luminosity moves upward. The effect is similar to that of sand flowing in an hourglass: while the sand flows downward, the effect of this flow is felt at higher and higher sections of the hourglass.

Uman, M. A. All about Lightning. *New York: Dover Publications, 1986, pp. 73–79.*

344. Outdoor Electric Field

A person standing outdoors forms an excellent grounded conductor, and her skin is basically an equipotential surface, like the surface of any conductor. The voltage on her skin is everywhere nearly the same value and approximately equal to the voltage of the ground. In some cases, a small atmospheric electrical current may flow through her body, but its value is smaller than the normal "biological currents." In most cases, the large mismatch in impedance between a person's body and the atmosphere plus the very small atmospheric current density prevents large currents even when the potential difference is 100 kilovolts!

Bering, E. A. III; A. A. Few; and J. R. Benbrook. "The Global Electric Circuit." Physics Today *51 (1998): 24–30.*

Dolezaler, H. "Atmospheric Electric Field Is Too Small for Humans to Feel." Physics Today *52 (1999): 15–16.*

345. Negative Charge of the Earth

The negative charge of the Earth seems to be related to the fact that the lower part of a thundercloud is predominantly negative, and about 85 percent of lightning bolts carry negative charge to earth. A mature thundercloud is tripolar, with a main negatively charged region at a height of about 6 kilometers sandwiched between two positively charged regions. On the scale of the global circuit there is a nearly constant potential difference of 300,000 volts between the negatively charged Earth and the upper atmosphere. A fair-weather leakage current of about 2,000 amperes constantly transfers positive charge from the upper atmosphere to the earth. It appears that thunderstorms in the tropics, particularly in the Amazon basin, which transfer large amounts of negative charge to the ground, are the dominant agent in recharging the global circuit.

Williams, E. R. "The Electrification of Thunderstorms." Scientific American *259 (1988): 88–89.*

346. Peak in the Global Electric Field

Universal time of 1900 corresponds to midafternoon in the Amazon basin, a region of particularly violent thunderstorm activity. The shape of the daily variation in the global electric field follows global thunderstorm activity. The thunderstorm rate is not a constant because continents are irregularly distributed in longitude, and thunderstorms occur primarily over land.

Bering, E. A. III; A. A. Few; and J. R. Benbrook. "The Global Electric Circuit." Physics Today *51 (1998): 24–30.*

347. Radio Reception Range

AM radio waves go farther at night. The phenomenon results from the existence of several ionized layers in the atmosphere at heights ranging from about 30 miles to more than 100 miles. The lower layers either disappear or diminish at night because the ionization of the molecules on the lower side of the ionosphere is reduced in the absence of sunlight. This raises the reflecting levels for both AM and shortwave signals and allows them to travel farther around the curve of the earth.

348. Car Radio Reception

The relatively low frequencies (535 kilohertz to 1605 kilohertz) used for AM (amplitude modulation) radio transmission correspond to wavelengths of 200 to 500 meters. Electromagnetic waves of such length are easily absorbed by large objects. This is why a pocket radio is unsatisfactory when used in a steel frame building. FM (frequency modulation) radio transmission, on the other hand, makes use of very high frequencies (VHF), ranging from 88 to 108 megahertz. These correspond to wavelengths of about 3 meters. In fact, the FM radio band is situated right in the frequency gap between television channels 6 and 7. Signals in this frequency range, including television signals, are not absorbed by large objects. For this reason they are reflected from them and scattered in all directions. Occasionally, both direct and reflected signals from the same station may be received at the same time. On TV this causes "ghost images," and on FM stereo it results in distortion or noise. However, barring such events, FM reception is not affected seriously by large objects, particularly in strong signal areas.

349. Magnetic Bathtubs

In the United States, if we take a compass needle and pivot it so that its ends can move up or down, we'll see that the north end will dip about 60 degrees to 70 degrees from the horizontal. One look at the globe will convince us that the north end is simply pointing along the shortest route through the earth to the magnetic pole in northeastern Canada. Similarly, the magnetic domains in stationary iron objects turn around until they line up with their north-seeking ends pointing 60 degrees to 70 degrees downward, with the south-seeking ends pointing directly opposite. The combined effect of millions of such magnetic domains all pointing in the same direction produces a magnetic north pole at the bottom and a magnetic south pole at the top of the object.

350. The Bathtub Vortex

We can expect the bathtub vortex effect to occur if the Earth's rotation is the dominant influence. As seen from the Northern Hemisphere the Earth's rotation is counterclockwise, whereas it is clockwise as seen from the Southern Hemisphere. The effect then could be regarded as one of the many manifestations of the Coriolis acceleration, which causes objects moving over the surface of the Earth to drift to the right north of the Equator and to the left south of it. However, the ratio of the Coriolis acceleration to the gravitational acceleration is roughly $2\omega v/g$, where ω is the angular velocity of the Earth. The ratio is on the order of 10^{-5} for water speed of, say, 1 m/s. Hence the relative importance of the Coriolis force in bathtubs and washbowls is negligible.

In practice the time involved is so short and competing factors (such as the long-term memory of the water for the direction in which it swirled and the asymmetries in the shape of the container) so numerous that any Coriolis effects will be swamped. Nevertheless, the effects do show up very clearly when the experimenters use highly symmetric hemispherical bowls and let the water rest for one or two days to eliminate any motions remaining from the filling process.

Shapiro, A. "Bathtub Vortex." Nature *196 (1962): 1080.*

Trefethen, L. M.; R. W. Bilger; P. T. Fink; R. E. Luxton; and R. I. Tanner. "The Bathtub Vortex in the Southern Hemisphere." Nature *207 (1965): 1084.*

351. Gravity Near a Mountain

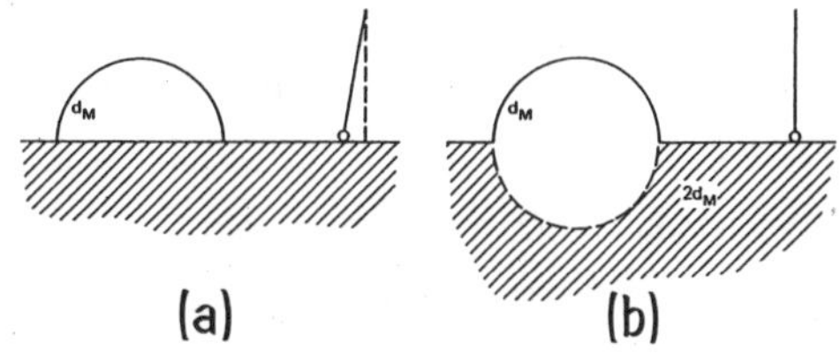

You might think that a mountain range could be represented by a long half cylinder of density d_m lying on a flat plane (see diagram, [a]). This model, however, predicts angles of deflection of a plumb bob that are much larger than what is actually observed. Suppose instead that the mountain range can be represented by a long cylinder of density d_m *floating*

in a fluid of density 2 d_m (see diagram, [b]). The plumb bob deflection due to the mountain range is zero in this model. This second model makes good physical sense: the mass contained in the top and bottom halves of the cylinder is exactly the same as the mass of the earth that would be found in the bottom half of the cylinder if the mountain range weren't there. The success of this model has convinced geologists that mountains, and also continents, float on the underlying mantle rock.

352. Gravity inside the Earth

No. The simple linear relationship does not hold inside the real Earth. In fact, the gravitational field strength $g(r)$ exceeds its surface value throughout most of the interior volume because of the nonuniform density of the Earth. The average density of the innermost part of the Earth is about twice the average density of the entire Earth. Pressures and temperatures increase so much in the interior that the center of the Earth is as hot as the surface of the sun!

Hodges, L. "Gravitational Field Strength inside the Earth." American Journal of Physics *59 (1991): 954–956.*

353. Why Is Gravitational Acceleration Larger at the Poles?

The variation of g between the polar and equatorial values is about 5.2 cm/s^2. Most of it, specifically 3.4 cm/s^2, is due to the centrifugal effects—the fact that because of its rotation, the Earth is not an inertial frame of reference. The remainder is 1.8 cm/s^2. Only two-thirds of this remainder, or 1.2 cm/s^2, could be due to changes of the polar radius from that of a sphere of equal volume. The reason is rather technical. It turns out that in a small ellipsoidal flattening of a sphere, keeping the volume constant, the polar radius is shortened by twice as much as the equatorial radius is increased. A calculation then shows that only 0.44 cm/s^2, approximately one-third of the 1.2 cm/s^2 that has to be accounted for, can be attributed to the flattening of the Earth. Most of it will come from the fact that the Earth's density is not uniform but is larger near the center of the Earth.

Iona, M. "Why Is g Larger at the Poles?" American Journal of Physics *46 (1978): 790.*

354. The Green Flash

The earth's atmosphere behaves like a giant prism. It refracts (bends) the components of sunlight, the shorter wavelengths (violets, blues) being bent more than the longer ones (reds, oranges, yellows). The amount of this angular dispersion of white sunlight increases when sunlight passes through more air before reaching the observer, at sunset and sunrise.

The diagram illustrates how the shorter wavelengths deviate more sharply and appear to come from points higher in the sky than the longer wavelengths. Note: The eye-brain system assumes that a light ray originates from a point lying on the tangent to the path of the ray. (The letters in the diagram refer to the colors of the various components.) Thus the sunlight spectrum has violets on top and reds on the bottom. If a fair portion of the solar disk is visible above the horizon, the light rays from its various parts overlap and the spectrum cannot be seen; but as the sun sets, the colors of its spectrum should theoretically vanish one by one, the red rays first and the violet rays last. However, two other atmospheric effects must be taken into account: (1) the absorption of light, due mostly to water vapor, oxygen, and ozone, which screens out mostly the orangish and yellowish light; and (2) the scattering of light, with the shorter wavelengths (violets and blues) mostly affected. The only relatively unscathed color is green, which is what reaches our eyes. At high altitudes, where the air is usually clearer, the shorter wavelengths may still come through, and the flash can be blue or violet instead of green.

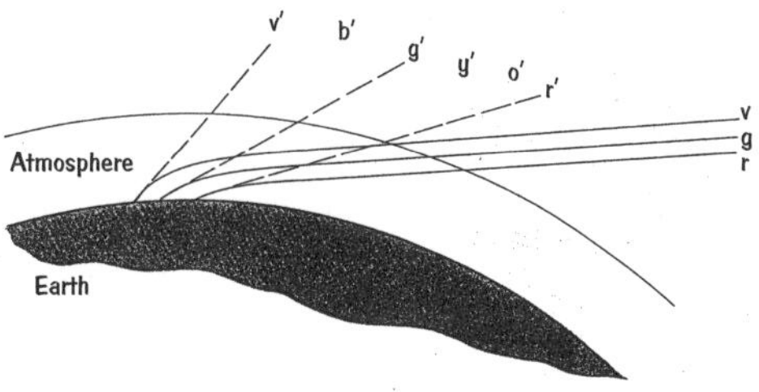

The flash lasts longer if the sun sinks relatively slowly—in winter at any one place (since the sun's apparent path makes the smallest angle with the horizon then), and at all times nearer the poles. At Hammerfest, Norway (latitude 79 °N), the flash at midsummer may last fourteen minutes: seven minutes during sunset and another seven minutes during sunrise—which follows immediately!

Connell, D. J. K. "The Green Flash." Scientific American *202 (1960): 112.*

Shaw, G. "Observations and Theoretical Reconstruction of the Green Flash." Pure and Applied Geophysics *102 (1973): 223.*

*355. Meandering Rivers

There are three different ways to look at the origin of meanders. The first is the mechanical model. Assume that a small bend of a river comes into being due to some minor irregularity of the terrain. The centrifugal force that arises as the water goes around the bend tends to fling the water outward toward the concave bank. Because the water at the top surface of the river is slowed less by the friction of the riverbed, it moves across the stream toward the concave bank and is replaced from below by water that moves across the bottom of the stream in the opposite direction (see the diagram). The concave bank is scoured by the downward current and eventually eroded, thus increasing the sharpness of the bend. This whole process throws the river into a path that traverses the hill rather than coursing straight down. Eventually, however, gravity pulls the river around into a downhill path, creating an opposite bend. Thus the process continues.

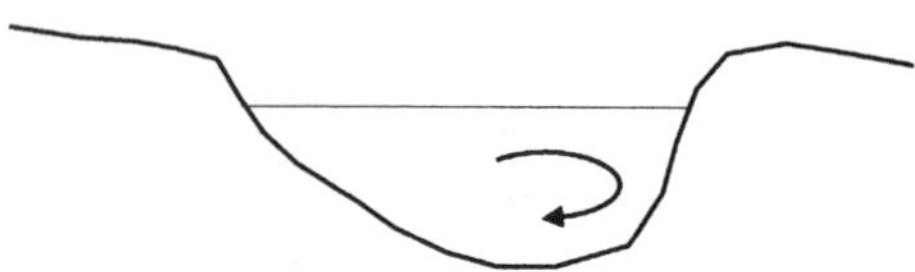

Looking at meanders from a different point of view, they appear to be the orm in which a river does the least amount of work in turning. Clearly, work is required to change the direction of a flowing liquid. The work is minimized if the shape of a river has the smallest total variation of the changes of direction. This property can be demonstrated by bending a thin strip of spring steel into various configurations by holding the strip firmly at two points and allowing the length between the fixed points to assume an unconstrained shape (see the diagram). The strip will assume a shape in which the direction changes as little as possible. This minimizes the total work of bending, since the work done in each element of length is proportional to the square of its angular deflection. The bends are not circular arcs, parabolic arcs, or sine curves; they are special functions known as elliptic integrals.

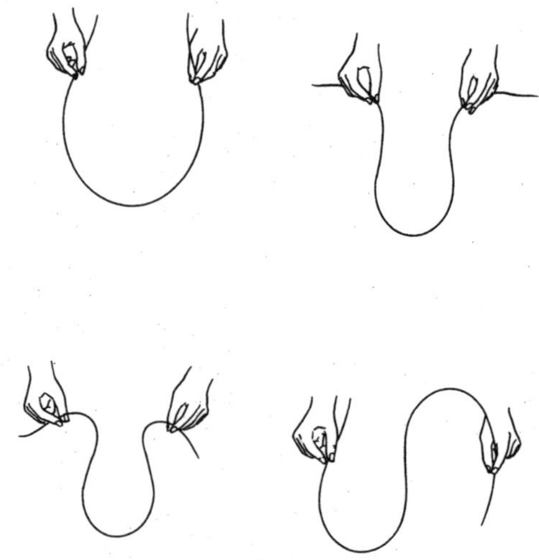

The third model for meanders comes from analyzing the course of a river in terms of randomness and probability. It is possible to prove that any line of fixed length that stretches between two fixed points is likely to follow a meander. The proof consists of generating random walks or paths in which a moving point can strike off in a direction determined by some random process (e.g., the throw of a die or the sequence of a table of random numbers) as it journeys between two fixed points in a specified number of steps. The most probable path for such a moving point is a serpentine pattern, with proportions similar to those found for rivers.

Einstein, A. "The Cause of the Formation of Meanders in the Courses of Rivers and the So-Called Beer's Law." In Essays in Science. *New York: Philosophical Library (1955), pp. 85–91.*

Leopold, L. B., and W. B. Langbein. "River Meanders." Scientific American *214 (1966): 60.*

*356. Energy from Our Surroundings

The heat reservoir is simply the night sky! A parabolic reflector with a black-painted body (i.e., "black" in the infrared, because black in the visible does not usually mean the same thing) at the focus, pointed at the night sky, will radiate in the infrared at the ambient temperature of, say, 300 K. It will receive little radiation from the night sky, which can be adequately regarded as blackbody radiation at 285 K. As a result, the temperature of the object at the focus will drop and, if thermally isolated from its surroundings, its temperature would eventually approach 285 K. We can then use the resulting temperature difference to run a heat engine or extract energy in other ways (such as by thermoelectric effects).

Ellis, G. F. R. "Utilization of Low-Grade Thermal Energy by Using the Clear Night Sky as a Heat Sink." American Journal of Physics *47 (1979): 1010.*

*357. Temperature of the Earth

There was no mistake, but we did leave something out. The equilibrium temperature T was found using this equation: flux absorbed = flux emitted, or $S(1 - A)\pi R^2 = \sigma T^4\ (4\pi R^2)$, where $S = 1.4 \times 10^6\ erg\ cm^{-2}\ s^{-1}$ is the solar constant and $A = 0.3$ is the typical value of the earth's reflectivity or albedo. The energy absorbed is mainly in the visible part of the spectrum, while the energy radiated back into space is mostly in the infrared. And here is the crux of the problem: we left out the greenhouse effect! While the atmosphere is very transparent at ordinary visible wavelengths, it is not as transparent in the infrared. When we calculate how much opacity is provided by the infrared-absorbing gases such as water vapor, carbon dioxide, methane, and chlorofluorocarbons (CFCs), we come out with the right answer.

Sagan, C. "Croesus and Cassandra: Policy Response to Global Warming." American Journal of Physics *58 (1990): 721.*

*358. The Greenhouse Effect

Both views are reasonable depending on the specific conditions. For a solar collector such as a greenhouse or the atmosphere of the earth, the power transferred by convection (in watts/ m^2) is $h\Delta T$, where ΔT is the difference between the outdoor temperature and the operating temperature of the collector, and h is a proportionality constant that increases with wind speed. The power emitted by radiation is approximately equal to $4\sigma T^3 \times T$, where σ is the Stefan-Boltzmann constant. When the air is still, the radiation loss is slightly larger, but when the wind is blowing at about 7 *m/s*, a typical value used by heating engineers for calculating winter heat losses, the convection loss increases to about five times the loss due to radiation.

If the collector is covered with a material that is transparent to infrared, then convection losses are halved (for still air), but the radiation loss is unchanged and becomes the dominant factor. However, radiation could be trapped effectively if we used a material that transmits visible light and reflects infrared. Such materials exist but usually are expensive.

Young, M. "Solar Energy: The Physics of the Greenhouse Effect." Applied Optics *14 (1975): 1503.*

———. *"Questions Students Ask: The Greenhouse Effect."* Physics Teacher *21 (1983): 194.*

*359. Measuring the Earth

The method requires a clear view of the sunset from a beach overlooking an ocean or a large lake. (Note: For eye safety reasons, it is best to avoid gazing at the sun's disk until it is mostly below the horizon.) Lie down so your eye is virtually at the water's level. Wait for (and note on your watch) the very instant at which *the last ray* of the sun suddenly shrinks (horizontally) and disappears. Stand up right away, and again note the time of the final ray from the *second* sunset. By subtraction, find the time elapsed between the two events (typically 10 to 20 seconds). Now (a) divide the eye height h (in meters) by the square of the elapsed time t and then (b) multiply the result by 378. The result is your own estimate of the earth's radius, expressed in thousands of kilometers. You may even desire to use the more complete approximate expression for the Earth's radius $R \approx h/(\omega^2 \cos^2\theta\, t^2)$, where θ is your latitude, with the factor 378 being the value for ω2 at the Equator in the given units.

Rawlins, D. "Doubling Your Sunsets, or How Anyone Can Measure the Earth's Size with Wristwatch and Meterstick." American Journal of Physics *47 (1979): 126.*

Walker, J. "How to Measure the Size of the Earth with only a Foot Rule or a Stopwatch." Scientific American *240 (1979): 172.*

Chapter 12 Across the Universe

360. Visibility of Satellites

An artificial earth satellite can be seen only if it is above the horizon and the sun is illuminating it from below the horizon. When the sun is in the sky it shines too brightly to allow you to see the satellite. Since many satellites, including those used for reconnaissance purposes, have near-polar orbits, an easy way to spot a satellite is to search the night sky near the North Star.

361. A Dying Satellite

By coincidence, the orbit of the nearest possible satellite—one just grazing the atmosphere—is very nearly 90 minutes. Because 90 minutes is exactly one-sixteenth of a day, the earth rotating underneath, after 24 hours, will bring the satellite back to almost the same spot in the heavens.

362. Cape Canaveral

Cape Canaveral was selected because of the land-free ocean extending 5,000 miles to the coast of South Africa. This fact is important because it makes it possible for the first two stages of the three-stage rockets launched over the Atlantic to fall into the water with little chance that they will fall on populated areas. Similarly, with the space shuttle, the booster rockets need to parachute into the ocean to be picked up and reused.

Why choose an East Coast launching site instead of a West coast one? Earth's rotation provides the answer. A rocket on the ground at Cape Canaveral is being carried eastward at 910 miles per hour. This speed is calculated by dividing the distance around the Earth at the latitude of Cape Canaveral (28.5 °N)—21,800 miles—by 24 hours. A satellite in low circular orbit must move at 17,300 miles per hour. If it is already going 910 miles per hour on the ground, the additional velocity required is only about 16,400 miles per hour. Currently, the launch pad in French Guiana (5 °N) takes best advantage of the free eastward boost due to the earth's spin. The Baikonur Cosmodrome (45.9 °N), east of the Aral Sea in Kazakhstan, has the least favorable latitude. Recent launches from a ship at the Equator in the Pacific Ocean have been able to take maximum advantage of the earth's rotation.

363. Weightlessness in an Airplane

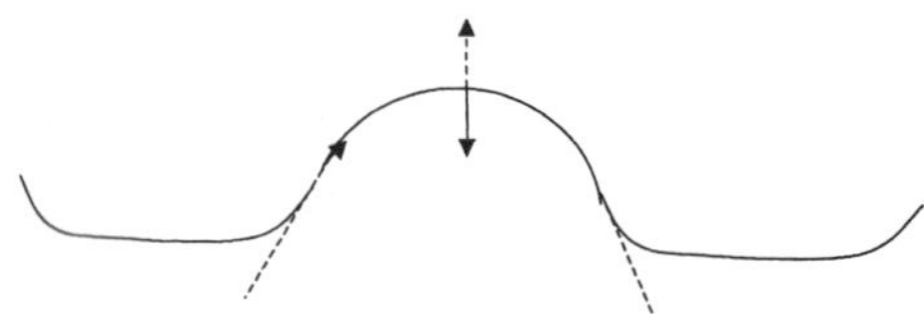

Weightlessness can be achieved when a plane flies a carefully controlled roller-coaster trajectory approximating answer (c). Near the top of each parabolic loop the centrifugal force (dashed arrow) that appears in the plane's frame of reference cancels out the gravitational attraction of the Earth (solid arrow), and the occupants become weightless. If this seems hard to believe, make a hole in the bottom of a can, fill it with water, and throw it at an angle to the ground. No water will be flowing out of the can while the can is in flight!

Weightlessness ends near the bottom of the loop, and for the next 40 to 50 seconds the plane climbs back up, pushing upward on the occupants with a force of about 2 g's (twice the force of gravity). On NASA flights training future astronauts, this roller-coaster ride may last up to an hour. One can understand why the old Boeing passenger jet used for this purpose has been dubbed the "Vomit Comet."

364. A Candle in Weightlessness

This question was investigated aboard the U.S. space station *Skylab* in 1973–1974. Contrary to popular descriptions, a candle can burn in zero gravity, albeit very slowly.

On earth, a candle continues to burn because of convection: warm air above the candle rises (being pushed up by the more dense air below), which causes more air to be pulled in at the bottom of the candle, thus resupplying it with oxygen. The rising convection current stretches the flame into its characteristic shape. In weightlessness there is no convection, so the flame will be roughly spherical. Combustion will occur only in a thin spherical shell, where the outward diffusing fuel vapors meet the inward diffusing oxygen. This restriction cuts the burning rate drastically. Here we are assuming that there are no air currents to provide more oxygen to the wick. That is not the case aboard space shuttles, where cabin fans constantly circulate air to cool the cockpit electronics. On a space shuttle, a candle would burn faster.

365. Boiling Water in Space

On earth we heat water mostly by convection. Heated water on the bottom of the kettle (near the heat source), being less dense, is displaced upward by the cold water on top, which sinks, gets heated, and rises again. These convection currents mix warm and cold water effectively.

There are no convection currents in a condition of weightlessness. Assuming that the side wall of the kettle has a very poor thermal conductivity and that no stirring device is present, the water on top is heated only by conduction—a slow process in water.

366. Maximum Range

Paradoxically, it is better to launch a spacecraft to reach as far out as possible into the Solar System when the Earth is closest to the Sun in its orbit—that is, at the perihelion. By choosing the perihelion date (about January 3), when the Earth is moving most rapidly in the Solar System, you would get the maximum possible boost from the Earth's orbital velocity.

367. Air Drag on Satellites

Initially, air drag can speed up a satellite! For a circular orbit, the total energy of a satellite of mass m is $E = -GMm/2r$, where r is the orbital radius. The potential energy is $2E$, while the kinetic energy is $-E$. Hence for each unit of energy "lost" due to the atmospheric drag, the satellite will "lose" two units of potential energy as it spirals down but will *gain* one unit of kinetic energy. This process cannot continue indefinitely. Gradually the drag force will become stronger and stronger while the gravitational force increases only slightly until drag is no longer a small perturbation but completely dominates the picture. Air drag will then act as a true braking force and slow down the satellite as it plunges to Earth.

Note that for the elliptical orbits the drag is greatest at perigee, where the speed and atmospheric density are both at maximum, and at minimum at apogee. Because of this difference, the orbit will become more nearly circular as it shrinks.

Berman, A. I. Space Flight. *Garden City, N.Y.: Doubleday, Anchor Press, 1979, pp. 85–88.*

Blitzer, L. "Satellite Orbit Paradox: A General View." American Journal of Physics *39 (1971): 882.*

368. Separation Anxiety

The launching rocket is generally larger than the satellite. As a result, it encounters more air resistance and slowly loses altitude. In doing so, the rocket converts some of its potential energy into increased kinetic energy—that is, greater speed. Thus the increased speed follows from the principle of the conservation of energy.

369. Changing the Orbit—Radial Kick

One might be tempted to answer that the orbit will elongate in the direction of the thrust. In fact, the orbit will elongate, but in a direction perpendicular to the kick, as shown in (c).

To get an insight into this counterintuitive result, compare the two orbits. Obeying the conservation of the angular momentum *mvr*, the maximum velocity will occur at the perigee. For orbit (b), v_{max} points horizontally to the right, and for orbit (c) vertically upward—that is, in the direction of the kick. Hence the radial thrust will produce orbit (c) since the maximum velocity must be in the same direction as the kick. Note that an inward radial thrust at the bottom of the original circular orbit would have the same effect.

Abelson, H.; A. diSessa; and L. Rudolph. "Velocity Space and the Geometry of Planetary Orbits." American Journal of Physics *43 (1975): 579.*

370. Changing the Orbit—Tangential Kick

As in the previous question, intuition may suggest that the orbit will elongate in the direction of the thrust. As before, the orbit will elongate, but in a direction perpendicular to the kick, as shown in (c).

Compare the two orbits. The maximum velocity will occur at the perigee. For orbit (b) v_{max} points vertically upward, and for orbit (c) horizontally to the left—that is, in the direction of the kick. Hence, the tangential thrust will produce orbit (c), since the maximum velocity must be in the same direction as the kick.

Abelson, H.; A. diSessa; and L. Rudolph. "Velocity Space and the Geometry of Planetary Orbits." American Journal of Physics *43 (1975): 579.*

371. Exhaust Velocities

Yes. This paradoxical fact can be understood by realizing that the exhaust gases always come out at the same velocity relative to the rocket, while the latter is constantly accelerating. Obviously at some point the rocket's forward velocity will exceed the gases' backward velocity, and relative to the ground the gases will start moving forward. Mathematically speaking, one can derive an equation for the velocity v of a rocket at any given time t as a function of the initial mass m_0 of the rocket, the mass m of the rocket at a time t, and the velocity v_{ex} of the exhaust gases with respect to the rocket. The equation is simply $v = v_{ex} \ln (m_0/m)$ for the ideal case. It is easy to see from this equation that as soon as the rocket has burned fuel to the point where $m_0/m > e$, v becomes greater than v_{ex}, and that, with respect to the ground, the exhaust gases travel in the same direction as the rocket.

372. Liftoff Position

The effect of accelerations on the human body varies depending on whether the astronaut is lying in the direction of the acceleration, so that her blood is forced from the head to the feet; or whether she is lying in a prone position, so that her head and heart are at the same relative level as far as the g forces are concerned. In a sitting-up position, loss of consciousness occurs at 4 to 8 g, depending on the duration and whether the astronauts are wearing anti-g suits. On the other hand, in a prone position the astronauts can tolerate up to 17 g for short periods of time without losing consciousness.

At liftoff, shuttle astronauts experience an acceleration of 1.6 g; 1 g is an acceleration in which the speed changes by 9.8 meters per second during each second. In British units, 1 g is equivalent to a uniform acceleration from 0 to 60 miles per hour in about 3 seconds. For comparison, a typical jet airliner accelerates at about 0.33 g down the runway before takeoff. The g forces vary as the shuttle is ascending but never exceed 3 g. Finally, 8.5 minutes into the flight, main-engine cutoff occurs, and in a split second the astronauts go from 3 g to weightlessness. For comparison, during most of reentry the g forces never get as high. The maximum is typically 1.5 g.

Mullane, R. M. Do Your Ears Pop in Space? and 500 Other Surprising Questions about Space Travel. *New York: John Wiley & Sons, 1997, pp. 53–54.*

373. Escaping from Earth?

Yes, it can escape. The total energy of a rocket of mass m and speed v on the surface of the earth of radius R is $\frac{1}{2}mv^2 - GMm/R$. The first term is the kinetic energy of the rocket, and the second term is its negative potential energy in the gravitational well of the Earth. To escape from the earth, the rocket must have enough kinetic energy so that its total energy is zero or positive—that is, $1/2mv^2 - GMm/R \geq 0$. This condition is independent of the direction of v, so it doesn't matter which way the rocket is pointing. If the total energy is zero, the rocket follows a parabolic trajectory.

In practice, for speeds less than 11.2 kilometers per second it is far more economical to use a horizontal launch. For one thing, if the flight path is toward the east, the effective speed of the rocket is increased by the Earth's surface speed at the latitude of launch. Secondly, the horizontal flight path gives the greatest possible angular momentum, which simplifies the problem of matching speeds with an orbiting vehicle, or with a planet such as Mars, moving in the same direction.

Interestingly, the minimum speed required to escape from the Earth-Sun system does depend on the launch angle relative to the Earth's orbital velocity. The optimum solution, given the minimum speed of 16.6 kilometers per second, is to launch along the direction of the Earth's motion. Note that this speed is a much lower value than the incorrect speed of 42 kilometers per second often found in textbooks for escaping from the Sun, starting at the distance of 1 A.U. When launching radially away from the Sun, the minimum escape speed is 52.8 kilometers per second.

Berman, A. I. Space Flight. *Garden City, N.Y.: Doubleday. Anchor Press, 1970, pp. 56–57.*

Diaz-Jimenez, A., and A. P. French. "A Note on 'Solar Escape Revisited.'" American Journal of Physics *85 (1988): 85–86*

Hendel, A. Z. "Solar Escape." American Journal of Physics *51 (1983): 746.*

374. Orbit Rendezvous

The forward burn will have precisely the opposite effect: it will increase the distance between the shuttle and the space station. Thrusting toward the target increases the shuttle's energy, which takes it into a higher orbit. This result can be seen for a circular orbit in the relationship between the total energy and the radial distance r, $E_{tot} = -GMm/2r$. But a higher orbit is associated with lower speeds, as we can see from $v^2 = GM/r$, so the shuttle will slow down. The correct procedure requires a series of maneuvers. You would begin with a braking burn, which decreases the shuttle's total energy and drops it into an elliptical orbit. This orbit, after circularization, is lower, and hence faster than the target's orbit. After coming ahead of the space station, you would reverse this series of maneuvers to move back up into the target's orbit and slow down.

Wolfson, R., and J. M. Pasachoff. Physics. *Boston: Little, Brown, 1987, pp. 191–192.*

375. Shooting for the Moon

Because of the effects of the Sun's gravity on the Moon's orbit, the inclination of its orbit relative to the Earth's orbital plane can vary by ± 5°9'. Combining this amount with the 23°28' tilt of the Earth's Equator to its orbital plane, the inclination of the Moon's orbit with respect to the Earth's Equator varies from 18°19' to 28°37', or about 281/2 degrees, the exact latitude of the Kennedy Space Center. This latitude permits NASA to launch spacecraft directly eastward, taking full advantage of the Earth's rotational speed, into orbits that lie almost exactly in the plane of the Moon's orbit. One wonders: Did Jules Verne know the orbital mechanics for a lunar probe?

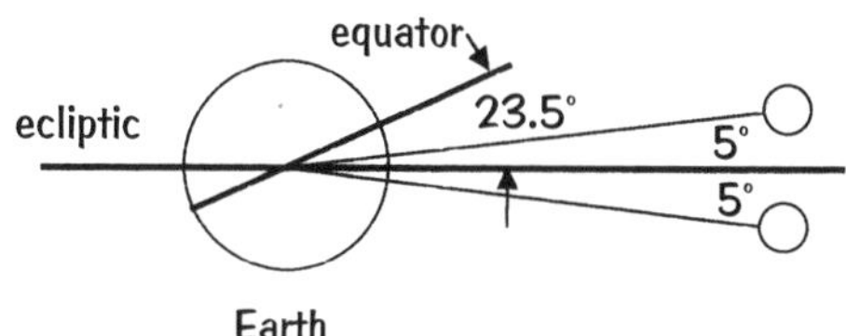

In contrast, the early Soviet lunar probes were launched from Tyuratam, east of the Aral Sea, which has a latitude of 45.6°. The best that could be accomplished from there was to launch into an orbit with an inclination of 45.6°, which is inclined about 17° to the Moon's orbit even under the best of circumstances. From there one must change to the Moon's orbital plane, a procedure that is very wasteful of fuel.

Lewis, J. S., and R. A. Lewis. Space Resources: Breaking the Bonds of Earth. *New York: Columbia University Press, 1987, pp. 132–137.*

376. Rocket Fuel Economy

Somewhat paradoxically, it is more economical to fire the upper stage when it is close to the ground than to fire it when its booster's apogee is reached. One gets the greatest benefit from a propellant when the upper stage is moving as fast as possible rather than when it is as far up as possible and moving very slowly. Mathematically, the change in kinetic energy is proportional to the speed—that is, $\Delta KE = mv\, \Delta v$.

Berman, A. I. Space Flight. *Garden City, N.Y.: Doubleday, Anchor Press, 1979, pp. 75–78.*

377. Speed of Earth

The earth moves fastest in winter and slowest when it's summer in the Northern Hemisphere. The Earth's path around the Sun is slightly elliptical, which means that the distance between the Earth and the Sun is constantly changing. Paradoxically for the inhabitants of the Northern Hemisphere, the Earth is closest to the Sun in winter and farthest away in summer. The perihelion, or closest point to the Sun (distance 1.471×10^8 km), is reached on January 2–5, depending on the year, and the aphelion, or the farthest point (distance 1.521×10^8 km), on July 3–6. Note, interestingly enough, that the Moon will appear a bit dimmer around the time of aphelion than around the time of perihelion. By Kepler's second law, the area swept out by the Earth's radius vector remains constant. To sweep out as large an area the Earth must move faster when it is close to the Sun, 30.3 kilometers per second at the perihelion and 28.8 kilometers per second at the aphelion.

378. Earth in Peril?

The Earth travels around the Sun at a speed of about 66,000 miles per hour. To drop inward and reach the Sun itself, the Earth would have to slow down drastically with respect to the Sun by accelerating the nearly 66,000 miles per hour in the direction opposite to its present motion. It is far easier to escape from the Sun completely than it is to get to the Sun.

379. The Late Planet Earth

The trajectory of the Earth falling into the Sun can be regarded as one side of a very skinny ellipse with a semimajor axis equal to 0.5 A.U. Using Kepler's third law, $T^2 = a^3$, the falling time is half the new period—that is, $T = 1/2\,(0.5)^{1.5}$ years or 64.6 days

380. Brightness of Earth

As Venus revolves around the Sun within the Earth's orbit, its sunlit hemisphere is presented to the Earth in varying amounts. It shows its full phase at the time of superior conjunction, the quarter phase on the average near elongations, and the new phase at inferior conjunction. Paradoxically, Venus is at its brightest not when it is nearest the Earth (its new phase), but in its crescent phase (about five weeks before and after the new phase). On the other hand, the Earth, being farther away from the Sun than Venus, presents all of its illuminated hemisphere toward Venus when the two planets are closest.

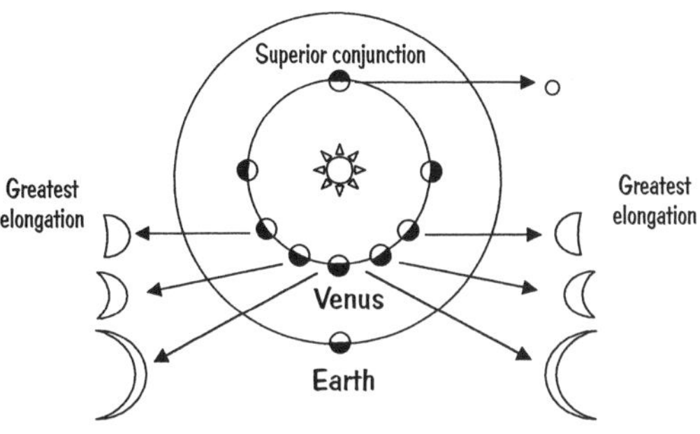

381. Meteor Frequency

The morning side of Earth is struck by both the meteors it encounters and those it overtakes, while the evening side is hit only by those meteors that gain on the Earth, as shown in the diagram.

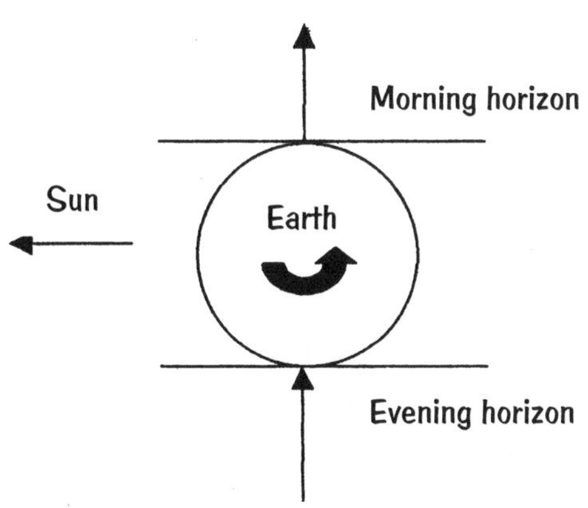

382. Slowly Rotating Earth

If the relationship displayed by the data obtained from the other planets held for the Earth, it would be rotating in 15.5 hours rather than 24 hours. However, over the ages the Earth's rotation has been slowed down by the tidal effects of the Moon. Other planets, Mars, Jupiter, Saturn, Uranus, and Neptune, do not have any satellites as large in relation to themselves as the Moon is in relation to the Earth. Therefore they have not suffered a comparable slowing effect.

The Moon itself suffers a slowing effect even greater than that sustained by the Earth. While the Earth is affected by the Moon's gravity, the Moon is affected by the Earth's 81-fold greater gravitational field. The Moon's rotation has been slowed to a complete standstill with respect to Earth, so that the same side always faces us. Its rotation with respect to the Sun, however, has not stopped. Its solar day is about 29.5 Earth days, which is equal to the time interval between two consecutive full moons.

The period of revolution of Mercury has been drastically slowed by the Sun's tidal effects and is now equal to 58.65 days, two-thirds of the planet's orbital period of 87.97 days. Hence, Mercury is locked into a three-to-two spin-orbit coupling, meaning that the planet makes three complete rotations on its axis for every two complete orbits around the Sun. Venus has also been slowed down by the Sun and now takes 243 days to rotate (backward!) on its axis, which is close to its period of revolution about the Sun (225 days).

383. Can the Sun Steal the Moon?

The sun gives the earth practically the same centripetal acceleration that it gives the Moon. Accelerations of bodies in a gravitational field are independent of their masses, so when we compare the Moon and the Earth, the only factor remaining is their relative distances from the Sun; but the difference is so small it can be neglected. Consequently, the paths of the Earth and the Moon around the Sun are being curved at the same rate, so their mutual distance remains practically the same.

384. Moon's Trajectory around the Sun

Yes. The moon's trajectory around the Earth is always concave relative to the Sun. The actual path looks like a regular thirteen-sided polygon whose corners have been gently rounded (see the diagram). To see why, suppose the Moon is directly between the Earth and the Sun. In this position, the Moon is being pulled in opposite directions by the gravitational forces of the Earth and the Sun. The ratio of the Sunward force to the Earthward force is approximately 2.2:1. Hence, that bit of the Moon's trajectory must be concave toward the Sun, and if it is, no other part of the trajectory could be convex to the Sun.

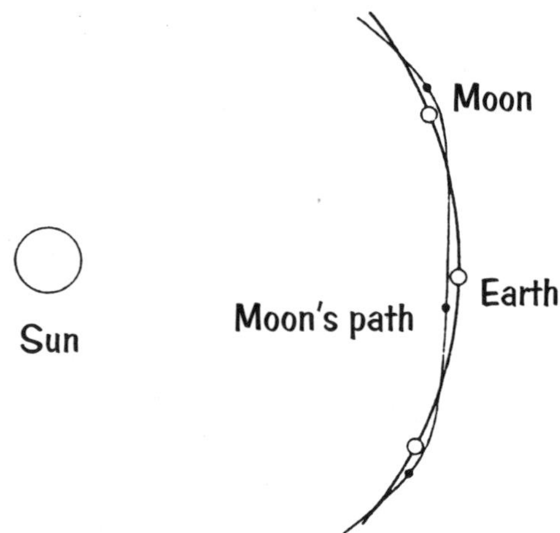

Purcell, E. M. "The Back of the Envelope." American Journal of Physics *52 (1984): 588.*

385. The Full Moon

The lunar surface is full of craters, mountain-walled plains, and other irregularities. These surface features cast long shadows when illuminated obliquely by the Sun, as during the first or the last quarter. The shadows make the surface appear darker than at full Moon, when the Sun shines directly from above over most of the lunar surface.

Note that due to the eccentricity of the Moon's orbit around the Earth, one full Moon is not equal to another! The distance to the Moon varies from as little as 354,340 kilometers (about 28 Earth diameters) to as much as 404,336 kilometers (about 32 Earth diameters), and accordingly, the light of the full Moon can vary by as much as 30 percent. Interestingly, the first-quarter Moon is about 20 percent brighter than the last-quarter moon.

Long, K. The Moon Book. *Boulder, Colo.: Johnson Books, 1988, pp. 39–42.*

386. The Moon Illusion—Luna Mendex

One reasonable explanation of the Moon illusion is the oldest one, going back at least to the second-century astronomer and geometer Ptolemy. It goes under the name of the apparent-distance theory and holds that the Moon low on the horizon appears to be farther away than the Moon high in the empty sky. The observer automatically takes the apparent distance into account, unconsciously applying the rule that, of two objects forming images of equal size, the more distant must be the larger (see the diagram).

Another reasonable and related explanation of the Moon illusion says that when the Moon is near the horizon, the ground and the horizon make the Moon appear relatively close. Since the Moon is changing its apparent position in depth while the light stimulus remains constant, the eye-brain size-distance mechanism changes its perceived size and makes the Moon appear very large.

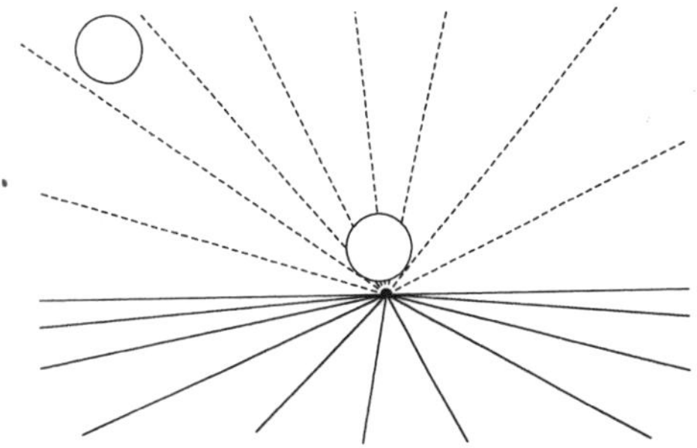

The history of the Moon illusion and details of alternative explanations can be found in the references listed. One should note that the Sun also suffers the same effect.

Hershenson, M. The Moon Illusion. *Hillsdale, N.J.: Lawrence Erlbaum Associates, 1989.*

Kaufman, L., and I. Rock. "The Moon Illusion." Scientific American *207 (1962): 120.*

Restle, F. "Moon Illusion Explained on the Basis of Relative Size." Science *167 (1970): 1092.*

387. Setting Constellations

Yes! See the explanation for the Moon illusion in the previous answer. The distances between the individual stars within a constellation appear to increase when the constellations are close to the horizon. This effect is particularly striking for the constellations Orion in winter and Cygnus in summer.

388. The Moon Upside Down?

The apparent orientation of the Moon's surface varies widely depending on the observer's latitude, and for a given latitude on the position of the Moon in the sky. Thus the lunar mountains (light areas) and maria (dark areas) can appear in vertical, horizontal, reversed, and all other intermediate positions, depending on where you are on Earth. If you take two observers along the same meridian—say, one in Boston and the other in Santiago, Chile—the observer in Chile will see the Moon exactly upside down compared to his friend in Boston only when the Moon is due south. At other times the relative orientation is more complicated.

389. How High the Moon?

When the ecliptic is low on the daylight side of the Earth, as in winter, it is correspondingly high on the dark side. Therefore the Moon is high on winter nights and low on summer ones, reaching its maximum height at full Moon, when it is directly opposite the Sun.

390. "Earthrise" on the Moon?

No. The Moon's rotation has become synchronized with its revolution about the Earth. As a result, the same hemisphere of the Moon is always turned toward the Earth. Superimposed on this is a "rocking motion," or libration, of the Moon that allows us to see, at one time or another, about 59 percent of the lunar surface, even though at any one time the most that can be seen of the surface is only 41 percent because the spherical shape of the moon hides the area close to the perimeter.

Thus, to an observer at a given site on the Moon, the Earth will basically appear at the same point in the sky, oscillating a bit about this position due to the libration. For example, from near the center of the visible lunar hemisphere the Earth will be visible directly overhead and will be seen to go through the phases the way the Moon does from the Earth.

391. Visibility of Mercury and Venus

Mercury and Venus follow orbits between the Sun and the Earth. As a result, to an observer watching the sky they are never far from the Sun, their maximum angle from the Sun, the so-called greatest elongation, being 28 degrees for Mercury and 48 degrees for Venus. Hence, when the Sun sets, Mercury and Venus are not far behind.

Several factors conspire to make Mercury rather difficult to see. Because its orbit is elliptical and has a 7-degree tilt to the plane of the ecliptic, the planet's greatest elongation can be as little as 18 degrees. Moreover, Mercury cannot be seen until it is at least 10 degrees away from the Sun. Consequently, even though Mercury may be as bright as some of the brightest stars, its periods of visibility are limited to a week or two three times a year in the evening and three times a year before sunrise.

Venus, which sometimes remains in the sky up to four hours after sunset, is in contrast quite easily visible in the sky. It is interesting that Venus, like the Moon, can on occasion be seen in full daylight, and warships have been known to fire at it, mistaking it for an enemy balloon.

392. Density of Earth

The gravitational field of the giant planets is high enough to attract and hold considerable atmosphere. The gases of such an atmosphere are low in density compared with the rocky main body of the planet, and their presence greatly reduces the density of the planet as a whole.

393. Rising in the West?

There are quite a few! One is the nearer and larger satellite of Mars, called Phobos, which revolves around Mars in 7 hours, 39 minutes. This period is less than a third of the rotation period of the parent planet. As a result, the easterly orbital motion of Phobos in the Martian sky far outweighs its apparent westerly motion caused by the rotation of Mars, thus making it rise in the west, gallop across the sky in only 5 1/2 hours, as viewed by an observer near the Martian equator, and set in the east.

Another object is the Sun as seen from Venus and Uranus. Viewed from the North Star, all planets revolve around the Sun counterclockwise and rotate around their axes also counterclockwise—that is, from west to east. Venus and Uranus are the only exceptions. Venus turns from east to west on its axis, and extremely slowly at that. Its day is equal to 243 Earth days. The retrograde rotation of Venus has, of course, the effect that the Sun rises very slowly in the west and sets just as slowly in the east. Uranus has its axis nearly parallel to the orbital plane, so the direction of the rising Sun changes by almost 180 degrees during one orbital year!

Even stranger is the behavior of the Sun as seen from Mercury's surface. When Mercury is near perihelion, the planet's rapid motion along its orbit outpaces its leisurely rotation about its axis. The Sun actually stops and moves backward (from west to east) for a few Earth days. In addition, Jupiter's outer four satellites, Saturn's moon Phoebe, and Neptune's moon Triton have retrograde orbits around their parent planets, which perhaps indicates that they are captured asteroids.

394. Taller Mountains on Mars

A mountain cannot rise higher than a certain critical height, which on the Earth is about 90,000 feet. Any greater height would increase the weight of the mountain to the point where its base would start turning into a liquid under such enormous pressures, thus causing the mountain to sink below the critical height. On the surface of Mars, the gravitation force per unit mass is less than on Earth; therefore the mountains are lighter, and they can reach greater heights.

*395. Going to Mars by Way of Venus!

Using the gravity-assist or slingshot method, the spacecraft undergoes an elastic collision with Venus in which there is no contact. Moving in the same general direction as Venus, the spacecraft approaches and leaves the planet with the same speed relative to the planet. Measured in the frame of reference of the Solar System, the spacecraft gains a small fraction of the planet's kinetic energy and in that frame emerges from the swingby with a higher speed, which sends it chasing after Mars. Round-trip time to Mars and back is about 500 days, more than a year shorter than in the transfer-ellipse method.

Roughly every 175 years the Jovian planets line up so that a single spacecraft can use the slingshot method to fly by all of them. *Voyagers 1* and *2*, launched in 1977, took advantage of such an opportunity to complete a grand tour of the giant planets between 1979 and 1989.

Berman, A. I. Space Flight. *New York: Doubleday, Anchor Press, 1979, pp. 167–172.*

Lewis, J. S., and R. A. Lewis. Space Resources: Breaking the Bonds of Earth. *New York: Columbia University Press, 1987, pp. 132–137.*

*396. Where Are You?

Spin a coin on the floor of your room. The coin will refuse to spin because, by conservation of angular momentum, the angular momentum vector of a spinning object tries to maintain its orientation in space, while the floor of the space station is rapidly changing its position in space.

*397. Was Galileo Right?

We have to be more precise. Do we mean the acceleration of a falling object relative to the center of Earth, or its acceleration with respect to the combined Earth-object center of mass? The latter location is called the barycenter. It is only the acceleration with respect to the barycenter that is independent of the mass of the object, as it is equal to the intensity of the terrestrial gravitational field at the center of mass of the object.

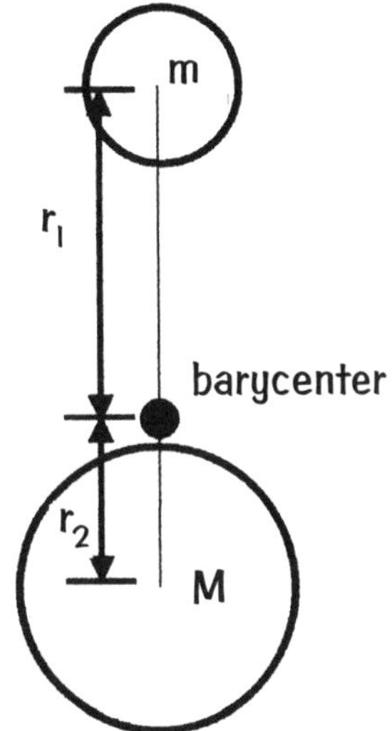

Of course, the Earth is simultaneously accelerating toward the falling object; hence the object's acceleration toward the center of the Earth is the sum of the accelerations of the object and the Earth. This effect *increases* with the mass of the object! Mathematically, $mr_1 = Mr_2$ or $(m + M)\ r_1 = M(r_1 + r_2)$, which can be transformed into $a_{m-M} = a_{cm}\ (1 + m/M)$, where a_{m-M} is the acceleration of the object with respect to the center of the Earth.

So perhaps Aristotle was right after all. Heavier bodies sometimes accelerate faster than lighter ones!

de la Vega, R. L. "Gravity Acceleration Is a Function of Mass." Physics Teacher *16 (1978): 292.*

Index